Arizona Ore Deposits

by the Arizona Bureau of Mines

with an introduction by Kerby Jackson

This work contains material that was originally published in 1938 by
the U.S. Dept. of Geological Survey.

This publication was created and published for the public benefit,
utilizing public funding and is within the Public Domain.

This edition is reprinted for educational purposes
and in accordance with all applicable Federal Laws.

Introduction Copyright 2014 by Kerby Jackson

Introduction

It has been decades since the Department of Interior released it's important publication "Some Ore Deposits of Arizona". First released in 1938, this important volume has been out of print and has been unavailable to the mining community since those days, with the exception of expensive original collector's copies and poorly produced digital editions.

It has often been said that "*gold is where you find it*", but even beginning prospectors understand that their chances for finding something of value in the earth or in the streams of the Golden West are dramatically increased by going back to those places where gold and other minerals were once mined by our forerunners. Despite this, much of the contemporary information on local mining history that is currently available is mostly a result of mere local folklore and persistent rumors of major strikes, the details and facts of which, have long been distorted. Long gone are the old timers and with them, the days of first hand knowledge of the mines of the area and how they operated. Also long gone are most of their notes, their assay reports, their mine maps and personal scrapbooks, along with most of the surveys and reports that were performed for them by private and government geologists. Even published books such as this one are often retired to the local landfill or backyard burn pile by the descendents of those old timers and disappear at an alarming rate. Despite the fact that we live in the so-called "Information Age" where information is supposedly only the push of a button on a keyboard away, true insight into mining properties remains illusive and hard to come by, even to those of us who seek out this sort of information as if our lives depend upon it. Without this type of information readily available to the average independent miner, there is little hope that our metal mining industry will ever recover.

This important volume and others like it, are being presented in their entirety again, in the hope that the average prospector will no longer stumble through the overgrown hills and the tailing strewn creeks without being well informed enough to have a chance to succeed at his ventures.

Kerby Jackson
Josephine County, Oregon
August 2015

PREFACE

The papers that make up this publication were written for presentation at the November, 1938, meeting of the American Institute of Mining and Metallurgical Engineers, at Tucson. Because they are reliable, up-to-date descriptions of many important Arizona ore deposits, the Arizona Bureau of Mines officials believed that they would be of much interest to persons concerned in the development of the state's mineral industries and requested permission to publish them as a bulletin. That permission was graciously granted by the officers of the Institute and the mining companies by whom some of the writers are employed.

The fifteen authors are eminent geologists and mining engineers of whom ten are not employees of the Bureau or the University. In no other way except for presentation at a great national meeting could so many authorities on the topics about which they write be induced to set forth their observations and conclusions, and the Bureau considers itself very fortunate to be able to make these papers available to Arizona mining men.

August 25, 1938   G. M. BUTLER, *Director*

TABLE OF CONTENTS

TABLE OF CONTENTS—*Continued*

PAGE

TABLE OF CONTENTS—*Continued*

TABLE OF CONTENTS—*Continued*

ILLUSTRATIONS

ILLUSTRATIONS—*Continued*

PAGE

SOME ARIZONA ORE DEPOSITS

PART I—GENERAL FEATURES[1]

By

B. S. BUTLER[2] AND ELDRED D. WILSON[3]

PHYSIOGRAPHIC SETTING

The principal ore deposits of Arizona are in the southern, central, and western portions of the state, which physiographically are part of the Basin and Range province, southwest of the Colorado Plateau (Pl. I).

The Basin and Range province is characterized by numerous subparallel mountain ranges separated by plains or valleys. Most of these ranges trend northwest to north, parallel to the margin of the Colorado Plateau; but in southeastern Arizona, southern New Mexico, and northeastern Sonora, they trend northward, transverse to the edge of the Plateau. The mountains rise abruptly from plains or valleys, the margins of which in many places are pediments cut on hard rock. Some of the plains form closed basins (bolsons, playas), but most of them are drained.

The Basin and Range province in Arizona is divisible into the Mountain Region and the Desert Region[4] (Pls. I and II). The Mountain Region forms a belt 60 to 100 miles wide that contains most of the large ore deposits. Its longest range measures about 55 miles, the widest 20 miles, and the highest peak more than 10,000 feet above sea level or 7,000 feet above adjacent valleys or plains. Broad plain-forming valleys are exceptional, but several with maximum widths of 20 to more than 30 miles appear in the southeastern portion.

In the Desert Region the mountains are relatively low, narrow, serrated, and steep, with sharply angular profiles. There are a few exceptionally large ranges of which the longest measures nearly 100 miles and the widest 18 miles. The highest peak is some 7,000 feet above sea level, and the maximum local relief amounts to about 5,400 feet. Broad and comparatively low desert plains form 50 to 75 per cent of the area.

---

[1] Paper prepared for the regional meeting of the A.I.M.&M.E. held at Tucson, Arizona, November 1-5, 1938.

[2] Professor of Geology, University of Arizona.

[3] Geologist, Arizona Bureau of Mines.

[4] F. L. Ransome, *Geology of the Globe Copper District, Arizona* (U.S. Geol. Survey Prof. Paper 12, 1903), pp. 10-16; *Copper Deposits of Ray and Miami, Arizona* (U.S. Geol. Survey Prof. Paper 115, 1919), p. 15.

In the Basin and Range province the mountain ranges and their intervening valleys or plains are largely of structural origin. Few of them appear to have been formed by erosion alone, and most of them reveal the structural or physiographic characteristics of mid-Tertiary and later tectonic masses which have been more or less modified by semiarid erosion. The topographic forms are due partly to the kind and structure of the rock masses and partly to the arid climate.

The mountain ranges are believed to represent blocks of the earth's crust which were elevated relative to the blocks that underlie the plains (see p. 20).

(see p. 20)

CLIMATE AND VEGETATION

The Mountain and Desert regions of Arizona are well known for their arid or semiarid climate and scanty vegetation.

TABLE 1.—TEMPERATURES AND ANNUAL PRECIPITATIONS AT SOME PLACES IN THE MOUNTAIN AND DESERT REGIONS OF ARIZONA.*

Place	Altitude (feet)	Mean temp.	Highest temp.	Lowest temp.	Precipitation (inches)
Mountain Region					
Crown King	6,000	......	......	....	32.42
Globe	3,525	62.7	110	10	16.70
Prescott	5,320	52.0	105	—12	18.52
Desert Region					
Douglas	3,930	61.1	111	— 7	13.68
Parker	350	69.3	127	9	5.35
Phoenix	1,108	69.7	118	16	7.90
Tucson	2,423	66.7	112	6	11.51
Yuma	141	71.7	119	22	3.10

* H. V. Smith, "Climate," *Arizona and Its Heritage* (Univ. of Ariz. Gen. Bull. No. 3, 1936), pp. 26-31.

TABLE 2.—MAJOR TYPES OF VEGETATION IN ARIZONA.*

Types	Altitude (feet)	Per cent of total
Forest		
Spruce-fir		
Yellow pine—Douglas fir	6,500-12,000	9
Piñon—juniper	5,000- 7,000	19
Chaparral	4,000- 5,500	7
Grassland	3,200- 6,500	23
Desert		
Sagebrush	2,500- 5,000	5
Creosote bush	137- 3,000	37

* H. L. Shantz, "Vegetation," *Arizona and Its Heritage* (Univ. of Ariz., Gen. Bull. No. 3, 1936), pp. 46-52.

Most of the rain falls during the summer and winter seasons, July-August and December-January. Both the rainfall and vegetation are strongly influenced by altitude.

STRATIGRAPHY

The columnar sections (Pl. III) must serve in the main to present the stratigraphy for this brief discussion.

PRE-CAMBRIAN

Rocks that have been classified as pre-Cambrian are widely exposed in the Southwest. As shown on the geologic maps, they constitute more than a third of the outcrops in the Arizona mountain ranges and are somewhat less prominent in southwestern New Mexico. These rocks comprise two major systems or types that are commonly termed Archean and Algonkian but in this summary will be referred to simply as older and younger pre-Cambrian.

Older pre-Cambrian.—In Central Arizona where the two systems occur separated by an unconformity, the older pre-Cambrian rocks include greenstone, rhyolite, shale, slate, grit and conglomerate, and the Mazatzal quartzite, intruded by diorite porphyry, porphyritic pyroxenite, granite, and granite porphyry. The intrusion of the igneous bodies culminated an intense crustal disturbance, the Mazatzal revolution,[5] characterized by folding and faulting of great magnitude. The granite, which includes the Bradshaw and Ruin granites, forms the largest batholiths exposed in the Southwest. Near large intrusive masses the older volcanic and sedimentary rocks grade into schist and have been collectively termed the Yavapai schist in the central region and the Pinal schist in southeastern Arizona.

It should here be stated, however, that schistose rock in southern Arizona is not necessarily of older pre-Cambrian age. Some sedimentary rocks probably as young as Mesozoic are highly schistose. In some descriptions Mesozoic and Paleozoic schists have been classed as Archean. Lack of schistosity is not a sure indication of later age, for in places the Pinal schist is but slightly schistose. Likewise, the coarse texture of igneous rocks is no certain basis for classification as older pre-Cambrian, since some of the post-Paleozoic intrusive rocks are strikingly similar to some of the older intrusive rocks.

Younger pre-Cambrian.—Following a period of erosion that cut deeply into some of the older pre-Cambrian batholithic bodies, there was deposited the Apache group which is largely confined to an area in central Arizona that extends north from Tucson to the Plateau, west into the Vekol Mountains, and east into the Little Dragoon Mountains. The Apache group, as now defined, consists of the Scanlan conglomerate at the base, succeeded by the Pioneer shale, Barnes conglomerate, Dripping Spring quartz-

---

[5] Eldred D. Wilson, *Pre-Cambrian Mazatzal Revolution in Central Arizona* (Geol. Soc. America, Proc. for 1936, 1937), pp. 112-13.

ite, Mescal limestone, and a flow of basalt. No fossils definitely indicative of age have been found in the Apache group. It is regarded as generally equivalent to the Unkar group (younger pre-Cambrian) of the Grand Canyon.

PALEOZOIC

General statement.—The Paleozoic consists of a series of prevailingly calcareous rocks, with some shale and sandstone, that were deposited in apparent conformity but include long intervals that are not represented by strata. The earlier classifications have been subdivided in certain areas, but for this summary it seems best to show only the larger units and correlations (Pls. II and III).

The Paleozoic section is much thicker in southeastern and northwestern Arizona than in the central part of the state.

Cambrian.—Resting with slight unconformity upon the Apache group or with profound unconformity upon older rocks is Cambrian sandstone or quartzite. It has been termed the Tapeats sandstone in northern and central Arizona; the Troy sandstone or quartzite in the Sierra Ancha, Globe, and Ray areas; the Coronado quartzite at Clifton; the Bolsa quartzite in southeastern Arizona; and the Bliss sandstone in southern New Mexico. Although not everywhere of exactly the same age, it is the basal formation of the Paleozoic section.

In southeastern Arizona the basal Paleozoic quartzite is overlain by a calcareous formation that contains much cherty material as well as sandstone and shale. This formation, which is of Upper Cambrian age, was originally called the Abrigo limestone by Ransome, but has been subdivided in some districts by A. A. Stoyanow.

Ordovician.—In the Clifton and Dos Cabezas districts of Arizona and in southwestern New Mexico, the basal Paleozoic quartzite is overlain by Ordovician beds, termed the Longfellow limestone at Clifton and the El Paso limestone in New Mexico.

Silurian.—In New Mexico the Silurian is represented by the Fusselman limestone, but no Silurian rocks have been recognized in Arizona.

Devonian.—Throughout most of southern Arizona, the Cambrian is succeeded directly by Devonian limestone and shale, known as the Martin limestone for most of southeastern Arizona, but as the Morenci formation in the Clifton district, and the Jerome formation at Jerome. In New Mexico it is called the Percha shale.

Carboniferous.—Conformably upon the Devonian is the Carboniferous which in most of the districts lacks clear-cut lithologic divisions.

The Carboniferous is known as the Tornado limestone in the Globe-Ray region and the Fierro limestone in the Silver City, New Mexico, area.

The lower or Mississippian portion in most of southeastern

Arizona is the Escabrosa limestone, with type locality at Bisbee. The Mississippian is represented by the Modoc limestone at Morenci and the Redwall limestone at Jerome.

The Pennsylvanian-Permian is called the Naco limestone, also with type locality at Bisbee. A. A. Stoyanow regards the upper part of the original Naco as Permian and has named it the Snyder Hill formation, with type locality near Tucson. The Pennsylvanian at Morenci is the Tule Springs limestone. The upper Carboniferous at Jerome is the Permian Supai formation.

Mesozoic

Early Mesozoic interval.—At the close of Paleozoic time, southern Arizona and adjoining parts of New Mexico were raised above the sea, and the break in their sedimentary record continues to the Cretaceous. In central Arizona there is a break in the stratigraphic record from the Permian to the Tertiary.

Some faulting and extensive erosion during the early Mesozoic interval has been shown by Ransome for the Bisbee district where almost the entire Paleozoic section was removed from the northeast side of the Dividend fault before Cretaceous sedimentation.

Igneous activity occurred in this interval, as shown north of Bisbee where the eroded surface of the Juniper Flat stock, intrusive into Paleozoic rocks, is overlain by the basal conglomerate of the Cretaceous. Pebbles of rock similar to the Juniper Flat intrusive, as well as of extrusive rocks, occur in this conglomerate as far distant as Tombstone, at least. As Cretaceous beds rest upon rocks of widely different ages throughout southeastern Arizona and southwestern New Mexico, the erosion during early Mesozoic was doubtless of irregular depth.

Cretaceous.—The basal member of the Cretaceous over wide areas is a conglomerate, local in derivation, that varies greatly in composition from place to place. In most areas it is succeeded by sandy and shaly beds with local limestone near the base at Tombstone and near the middle at Bisbee. The limestone, however, is very subordinate in amount.

Rocks of supposed Cretaceous age in parts of southern Arizona contain much volcanic material, but some of them are difficult to distinguish from Tertiary.

Because of deformation and incomplete exposures, the Cretaceous has been accurately measured in but few localities. Ransome found 5,000 feet of Comanchean beds in the Bisbee district, and the total section in some other areas is probably much thicker.

At Bisbee the Cretaceous Bisbee group has been subdivided into the Glance conglomerate. Morita formation, Mural limestone, and Cintura formation (Pl. III). In New Mexico the basal Sarten sandstone is overlain by the Colorado shales.

Laramide Interval

Although the events of late Cretaceous and early Tertiary time can not yet be precisely dated, their general order seems reasonably well established. Probably toward the close of Cretaceous and during early Tertiary time, the Southwest, in common with the Rocky Mountain region, underwent the structural deformation and igneous activity of the Laramide revolution. There occurred a great outpouring of volcanic material, accompanied or closely followed by intrusion of batholithic masses. It is thought that the intrusive bodies of the Globe, Miami, Ray, Superior, and Christmas areas represent this period, as probably do those of Morenci, Arizona, and of the Silver City-Santa Rita region, New Mexico. It has been suggested that the Bisbee intrusives are of this period, but Ransome and Trischka regard them as earlier.

CENOZOIC

During early or middle Tertiary the Southwest underwent a long period of erosion and sedimentation, represented by the Whitetail conglomerate of central Arizona. During this period of erosion, most of the region stood at a fairly high elevation above its water table, and conditions for supergene enrichment were exceptionally favorable.

After erosion uncovered some of the Laramide batholiths and stocks, they were again buried beneath volcanic material, and probably other intrusive bodies were injected. These later Tertiary lavas, which form series of great thickness (Pl. III), have in places been deeply eroded, re-exposing the earlier intrusive bodies and associated rocks. In few places, however, have any large intrusive masses that may have accompanied the later extrusives been extensively exposed.

The volcanic activity was accompanied and followed by upheaval of the present mountain ranges. During late Tertiary time, stream and lake sediments, in places thousands of feet thick, accumulated in the lower areas. This formation, though given various local names, is best known as the Gila conglomerate. Mild volcanic activity accompanied the sedimentation, and lava flows are locally present in and on top of the formation. Relative uplift of the ranges and subsidence of the intermont trough areas continued intermittently into Quaternary time, and the Gila formation is tilted, faulted, and locally folded. Where uplifted, it is trenched, and the material eroded from it and from other sources has been deposited as a relatively thin veneer of Quaternary terrace and stream alluvium.

IGNEOUS ACTIVITY

Igneous activity, which has already been mentioned in connection with the general stratigraphy, may be classified into the following general periods: Older pre-Cambrian, Younger pre-Cambrian (Apache), post-Jurassic, Laramide, and Tertiary.

PRE-CAMBRIAN

Older pre-Cambrian.—During early pre-Cambrian time, igneous activity was very extensive and probably continued over long periods, but it has not yet been subdivided and correlated beyond a few areas. Igneous rocks are present in most exposures of early pre-Cambrian in the Southwest. Batholithic masses of coarse-textured granite and bodies of dioritic to more basic rock occupy large areas.

In central Arizona greenstone (Pl. III) represents widespread eruptions and minor intrusions of intermediate to basic volcanic rocks during earliest pre-Cambrian time. This activity was followed by extensive eruptions and near-surface intrusions, prevailingly of rhyolitic composition, exemplified by the Red Rock rhyolite of central Arizona (Pl. III) and the Deception porphyry of Jerome. Rhyolite of similar aspect is present in the Pinal schist at Ray and at least as far west as the Vekol Range.

After a long period of time, measured by the accumulation of many thousands of feet of sedimentary rocks, there ensued the intrusive activity of the Mazatzal revolution, during which the most intense and widespread crustal disturbance recorded in the region culminated with the intrusion of the largest igneous bodies exposed in the Southwest, as exemplified by the Bradshaw and Ruin granites and other bodies of dioritic to mafic composition. Definite correlation of these and other so-called Archean intrusives throughout the Southwest must await age determinations based on radioactivity.

Younger pre-Cambrian.—The close of Apache time was marked by a flow of basalt, about 100 feet in maximum thickness.

Great sills of diabase invade the Apache strata, but the age of its intrusion is still somewhat in question. It has been regarded as Mesozoic (?) by Ransome,[6] Younger pre-Cambrian by Darton,[7] and post-Middle Cambrian by Short.[8]

MESOZOIC AND CENOZOIC

Post-Jurassic.—Igneous activity that took place between the Paleozoic and early Cretaceous is represented by the Juniper Flat stock in the Bisbee area. Although dated only within wide limits, this activity may be provisionally correlated with the post-Jurassic revolution of the Sierra Nevada.

Ransome, Trischka, and others regard the Sacramento stock at Bisbee as pre-Cretaceous, but this interpretation has been questioned.

Igneous activity of this period is probably represented in other areas where age relations are obscure.

Laramide.—Throughout the Southwest, a period of great igneous activity began in late Cretaceous time, as recorded by a thick series of flows and clastic volcanic rocks. In many localities, these eruptions were accompanied or closely followed by intru-

[6] F. L. Ransome, *Description of the Ray Quadrangle* (U.S. Geol. Survey Folio 217, 1923).

[7] N. H. Darton, *A Résumé of Arizona Geology* (Univ. of Ariz., Ariz. Bureau of Mines Bull. 119, 1925).

[8] Unpublished data.

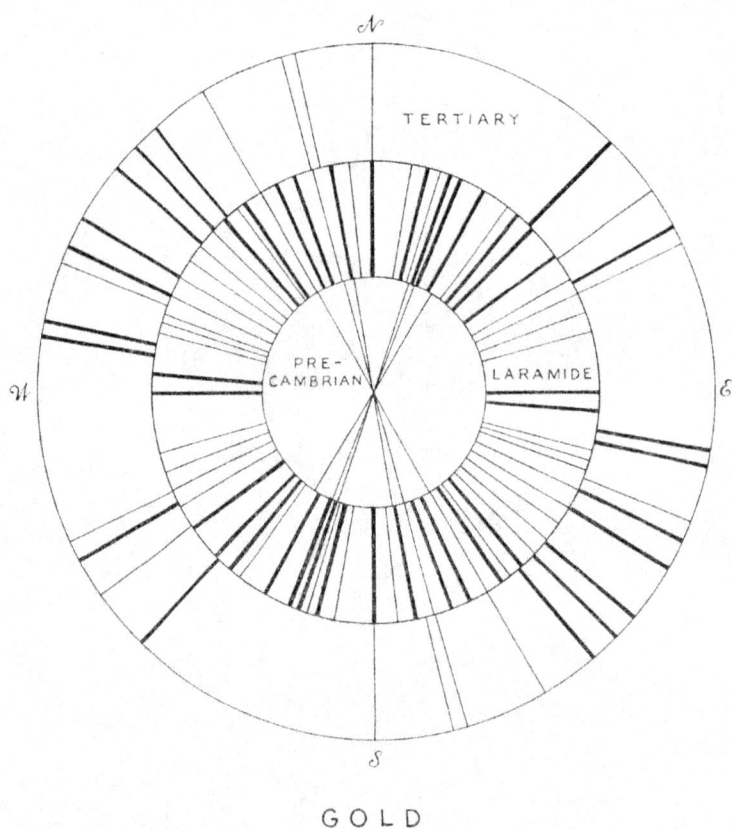

GOLD

Figure 1.—Diagram showing trends of principal gold-quartz veins in Arizona. Individual veins that have yielded $200,000 to $500,000 worth of gold are represented by light lines; more than $500,000 worth, by heavy lines.

sions of batholiths, stocks, and dikes, prevailingly of intermediate to granitic composition.

Although the age of most of these igneous rocks is not closely delimited, they may be grouped as of Laramide age. The oldest are late Cretaceous, whereas the youngest seem to extend into the Tertiary.

Tertiary and Quaternary.—In several districts, as Clifton, Silver City, Miami-Ray, and Bagdad, the Laramide intrusions were uncovered by erosion and later buried beneath a thick cover of volcanic rocks which in turn have been eroded to re-expose the

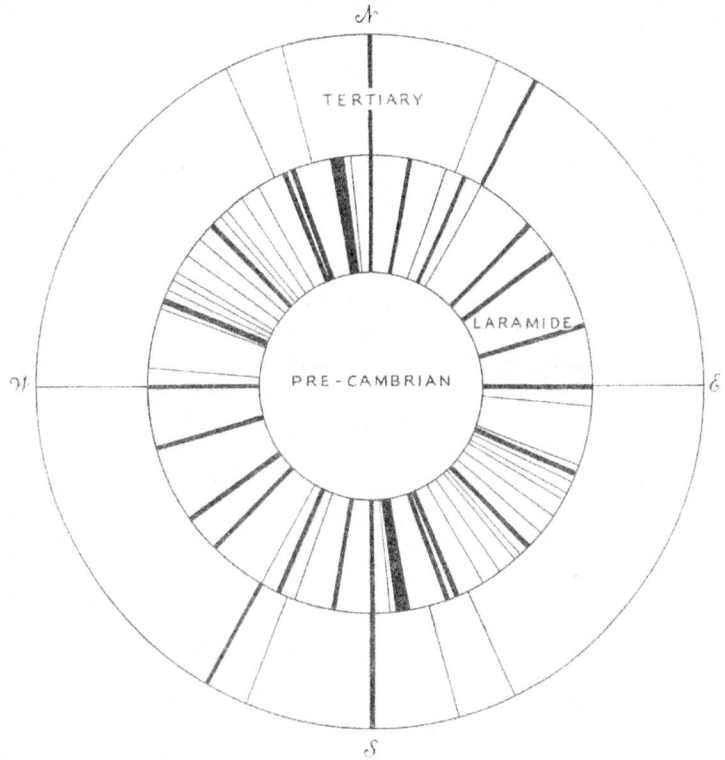

SILVER - LEAD

Figure 2.—Diagram showing trends of principal silver-lead deposits in Arizona. Deposits that have produced $50,000 to $500,000 worth of ore are shown by light lines; more than $500,000 worth, by heavy lines.

older rocks. In general, only the smaller of the intrusive bodies that accompanied this later volcanic activity have been exposed. Vulcanism continued, intermittently and probably with decreasing force, past the time of the early Indians. Probably it has not yet ceased.

STRUCTURE

The general features of structure are closely related to the periods of igneous activity, each of which was associated with a revolution or period of crustal disturbance.

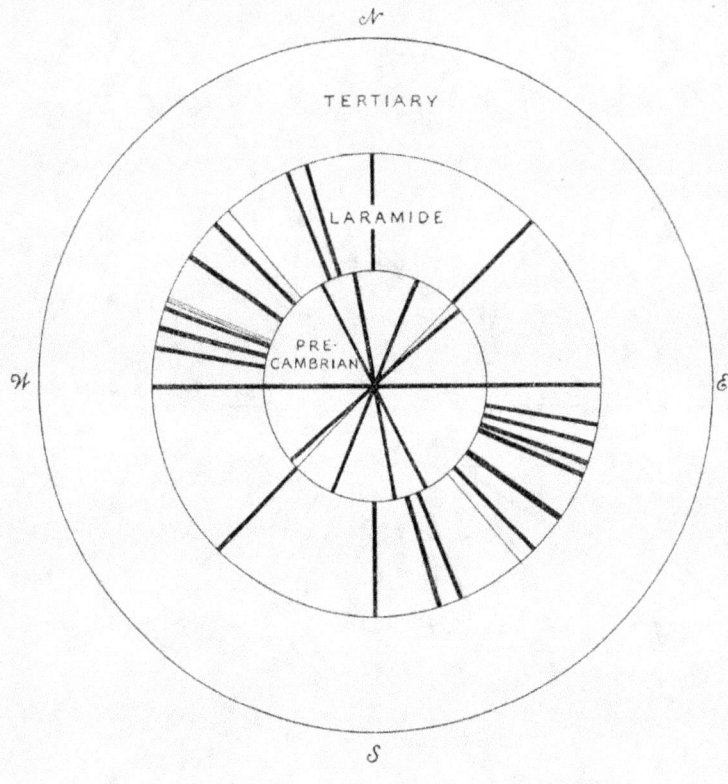

COPPER

Figure 3.—Diagram showing trends of principal copper deposits in Arizona. Deposits that have produced $50,000 to $500,000 worth of ore are shown by light lines; more than $500,000, by heavy lines.

PRE-CAMBRIAN

Older pre-Cambrian.—Intense deformation, with a predominance of northeast to northward-trending structures, is characteristic of the earlier pre-Cambrian rocks.

In central Arizona, the compressive stresses of the Mazatzal revolution (p. 11) gave rise to folds and faults of subparallel, northeast to northward trend, and faults and joints of northwest to westward trend. The thrust faults dip southeast to eastward and effected displacements of several miles.

Locally, as around large intrusive bodies, the trends show considerable variation, but these early structures probably exerted considerable influence upon the later tectonic patterns.

The close, deep folding gave rise to a broad mountain range that extended northeastward, diagonally across the state. As noted by Ransome,[9] parts of this range survived as a barrier through most of Paleozoic time. Stoyanow[10] has termed it Mazatzal Land. In its vicinity Paleozoic and Mesozoic sedimentation was thin compared with that in southeastern and northwestern Arizona.

The predominant trend of pre-Cambrian ore deposits in the Southwest is northeast to slightly west of north (Figs. 1, 2, and 3).

Younger pre-Cambrian.—So far as known, deformation in the Southwest during Younger pre-Cambrian time was comparatively slight and of little significance in connection with metallization. Mild faulting and doming preceded or accompanied the diabase intrusion, but its age is not definitely pre-Cambrian (p. 15).

MESOZOIC AND CENOZOIC

Post-Jurassic.—No pronounced deformation of Paleozoic age has been recognized, although regional uplift must have occurred at the time of the Appalachian revolution to provide the land masses from which the Triassic and Jurassic sediments of northern Arizona and New Mexico were derived. Faulting occurred during pre-Cretaceous, probably post-Jurassic time, as exemplified by the Dividend fault in the Bisbee district. This great fault strikes northwestward. Other structures of that period doubtless are present in the Southwest, but they have not been definitely dated.

Laramide and Tertiary.—Associated with the Laramide igneous activity (p. 15) was a great crustal disturbance, marked by folding and faulting in southern New Mexico, southeastern Arizona, and northwestern Arizona. Throughout the middle portion of the state, between longitude 111° 30' and 113° 30', the Paleozoic and later rocks show much faulting but little or no folding, although they are intruded by numerous igneous bodies which indicate the operation of Laramide mountain-making forces. In other words, strong Laramide folding and thrusting are apparent only in the areas of relatively thick sedimentation and are lacking in the positive area of Mazatzal Land.

The thrusting probably continued over a long period, for in places it affects Pliocene rocks.

There is yet much question as to the relation of thrust faults and intrusive rocks. Most intrusive bodies are later than the low-angle thrust faults, but the opposite may be true for some intrusive bodies.

The general trend of the thrust faults is northwestward with the mountain ranges.

Some structures trend across the mountain ranges. In the Superior-Miami-Globe and Morenci-Metcalf districts, for ex-

---

[9] F. L. Ransome, *Some Paleozoic Sections in Arizona and Their Correlation* (U.S. Geol. Survey Prof. Paper 98, 1916), pp. 165-66.

[10] A. A. Stoyanow, *Correlation of Arizona Paleozoic Formations* (Geol. Soc. America Bull., 1936), XLVII, 462.

ample, the intrusive bodies have a general northeastward trend. Likewise, the Laramide ore-bearing fissures in these and in many other districts strike northeastward, whereas the later Tertiary veins strike predominantly northwestward (Figs. 1, 2, and 3). Faults later than the ores strike north to northwestward. In the Empire Mountains, southeast of Tucson, are faults of thousands of feet of displacement that strike northward and eastward, transverse to the range, and possibly preceded the outlining of the Basin ranges.

These diverse structural trends may be the result of a system of stresses influenced by pre-Cambrian structures, but detailed mapping over a large region will be required before definite conclusions can be reached.

The present ranges of southern Arizona and southern New Mexico are generally regarded as originating from fault blocks, most of which were tilted. Erosion has removed the higher areas and deposited the debris in the depressions, building up the plains, while faulting continued intermittently. If this simple explanation applies to all of the ranges, it has been in progress over a long period, resulting in very different stages of physiographic development.

For many ranges the displacement along the border was not effected by a single fault, but by a zone of step faults. Many of these zones appear to be *en echelon*. That the faults are deep seated is indicated by the common occurrence of late basalt along margins of ranges.

MINERALIZATION

The ore deposits are closely associated with the igneous rocks, especially the intrusive rocks, and therefore, in time of formation, may be grouped with the periods of igneous activity, as Older pre-Cambrian, Younger pre-Cambrian, post-Jurassic, Laramide, Tertiary, and Recent.

PRE-CAMBRIAN

Older pre-Cambrian.—Deposits known to be of Older pre-Cambrian age are largely confined to the Jerome-Prescott and adjoining areas in Arizona and to central New Mexico where the Pecos deposit is best known.

Copper has been the most important metal produced from the Older pre-Cambrian deposits in Arizona. These copper deposits have also yielded much gold and silver, but the pre-Cambrian gold-quartz veins have produced comparatively little. Zinc is abundant in the Jerome deposit, and zinc and lead are most important at Pecos, New Mexico. It is probable that some, at least, of the tungsten and molybdenum deposits of western Arizona are of Older pre-Cambrian age.

Younger pre-Cambrian.—No important metal production has been made from deposits of known late pre-Cambrian age. The iron deposits of Canyon Creek, Fort Apache Indian Reservation, are probably of this age, as are the asbestos deposits of central Arizona.

Post-Jurassic.—Numerous deposits, especially those of western Arizona, may be post-Jurassic. Bisbee is the only large district that has been assigned to this age, and it may be later.

Laramide.—Deposits believed to be of Laramide age have been by far the most productive in the Southwest. Provisionally included in this class are Miami, Globe, Ray, Superior, Morenci, Ajo, Tombstone, Courtland, Silver Bell, Dos Cabezas, Cerbat Mountains, Twin Buttes, Magdalena, Santa Rita, Pinos Altos, and Tyrone.

Tertiary.—Deposits of middle to late Tertiary age have been valuable mainly for gold with some silver. The Oatman, Katherine, and Kofa districts of Arizona and the Mogollon district of New Mexico have been the most productive of this class. The Mammoth district of Arizona, which has produced molybdenum and vanadium, but mainly gold, may be classed with this group, as may also the Artillery Mountain manganese deposits.

Recent.—Recent deposits include gold placers[11] that are rather widely distributed over the region but, partly because of the difficulty of operation under desert conditions, have yielded relatively little gold.

RELATION OF ORE DEPOSITS TO IGNEOUS ROCKS

The lode deposits, except those of iron and manganese, are closely associated with igneous activity and especially with intrusive bodies.

The grouping of ore deposits around intrusive bodies, and especially around the tops of stocks, has long been recognized[12] and shown to be a general relation.

The association as now seen is dependent first on the depth below the surface at which the igneous body came to rest, and second, on the amount of erosion that has occurred since the deposits were formed. Figure 4 illustrates the relations to intrusive bodies of different depths, and it can readily be seen what would result from different depths of erosion.

It is probable that the different depth relations were present in each major period of igneous activity, and the deposits as we now see them are what is left after various stages of erosion.

The deposits have characteristics indicative of their depth of formation aside from their relation to the intrusive bodies. The deep deposits are in shear zones, and the ore minerals have largely formed as a replacement of the sheared rock. This contrasts with

[11] See *Ariz. Bureau of Mines Bull. 142,* (Univ. of Ariz., 1937), pp. 1-90.
[12] B. S. Butler, "Relation of Ore Deposits to Different Types of Intrusive Bodies in Utah," *Econ. Geol.,* X (1915), pp. 101-21. W. H. Emmons, *On the Mechanism of the Deposition of Certain Metalliferous Lode Systems Associated with Granitic Batholiths* (Am. Inst. of Min. & Met. Eng., Lindgren Volume, 1933), pp. 327-49.

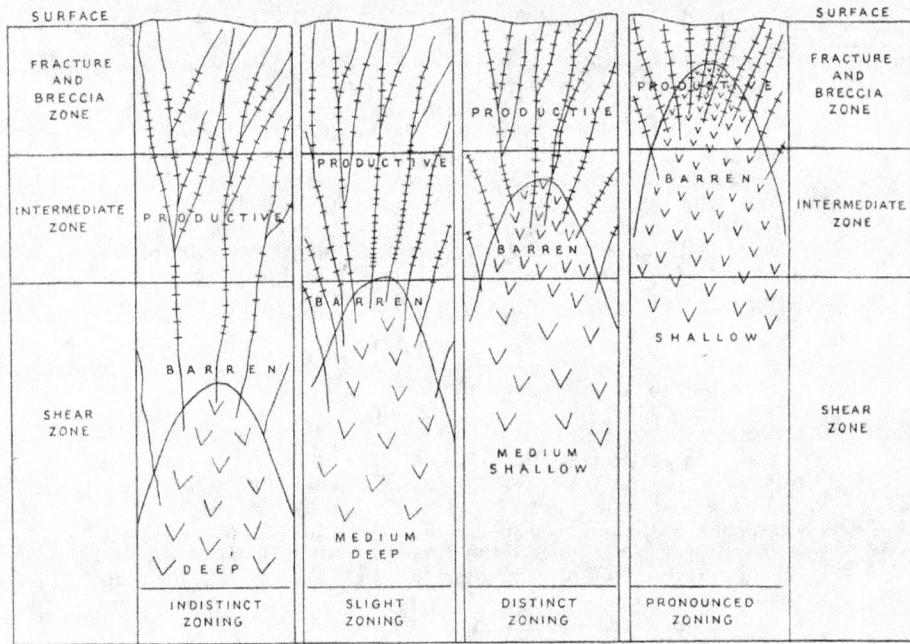

Figure 4.—Generalized and idealized diagram showing relation of ore deposits to associated stocks intruded at different depths.   (By B. S. Butler.)

the filling of fissures or the filling and replacement of breccia zones or breccia pipes in deposits formed nearer the surface. The minerals in the deep deposits are coarsely crystalline, whereas those formed near the surface are likely to be finer grained.

There is also a difference in the degree of separation of the metals. The deep deposits may have copper and zinc in the same lode, whereas nearer the surface the copper deposits are commonly in or very close to the intrusive body and contain little zinc or lead, and zinc and lead deposits are some distance from the intrusive bodies and are low in copper.

Different areas in the Southwest have been eroded to very different depths relative to the surface when intrusion took place at the different periods. The Older pre-Cambrian rocks have been deeply eroded, and over wide areas any near-surface deposits of that age have been removed. The pipelike deposits of Jerome and the shear deposits in adjacent areas suggest formation in a zone between deep and shallow.

In contrast with the deeply eroded pre-Cambrian, the later deposits are within and closely grouped about stocks, as exemplified at Bisbee and Morenci-Metcalf, Arizona, and at Santa Rita-Hanover, New Mexico. The more deeply eroded stocks seem to have had much of the lodes removed with the upper parts of the stocks. The Schultze granite stock, for example, between Superior and Miami, is nearly barren of mineral deposits within and around its central portion, but at either end where the roof of the stock plunges beneath the sedimentary cover valuable deposits are present.

The later deposits are in or associated with fissures, breccia zones, or breccia pipes, as contrasted with the shear-zone replacements of the deeper deposits, and there is more tendency for separation of the metals in districts where copper, zinc, and lead are present.

The Tertiary deposits are in part in Tertiary lavas, which indicates that their erosion has not been great. The veins fill fissures with only slight replacement, and vein minerals, especially the quartz, have the fine-grained, banded texture that suggests rapid change and rapid deposition. Characteristically, the deposits change rapidly with depth. Intrusive bodies with which the Tertiary deposits may be associated have not been exposed by erosion.

These near-surface types are well exemplified by the Oatman and Mammoth deposits of Arizona and the Mogollon deposits of New Mexico.

OXIDATION OF ORE BODIES

Alteration by weathering has been an important factor in the formation of the ore bodies. It has largely determined the value of many of the copper and some of the gold and silver deposits. The early production of copper in Arizona was from oxidized ores of simple metallurgy. Most of the disseminated deposits owe their value to changes resulting from weathering.

Weathering of the pre-Cambrian deposits began in that remote time and is now in progress. Many of the deposits have been subjected to more than one period of oxidation. The Jerome deposits were oxidized in pre-Cambrian time and then buried beneath later sedimentary rocks for a long period before faulting, and erosion exposed parts of them to renewed oxidation. Likewise, the Tertiary sedimentary rocks and lavas buried many of the deposits after oxidation was well advanced, and later erosion of these rocks again exposed them to renewed oxidation. Such was the Tertiary and later history of southwestern New Mexico, Morenci, Globe, and Bagdad, Arizona, and others.

In the different periods of oxidation the water table has been at different levels, which has doubtless been an important factor in the oxidation.

The aridity of the Southwest during Recent times has, in most of the districts, resulted in deep water tables, which are particularly favorable to weathering.

Copper deposits are of greatest importance in the region and have perhaps benefited most by weathering processes.

The results of weathering are dependent on the composition of the original deposits and the composition of the rocks enclosing the deposits, as well as on structure and climate.

Most of the primary deposits have been relatively rich in pyrite, which on oxidation produced sulphates of the metals and sulphuric acid, which in turn were favorable to the solution and transportation of copper. The extent of transportation depended largely on the enclosing rocks. The copper of deposits in limestone has moved relatively little, whereas that in siliceous rocks has undergone much movement.

Sulphide deposits oxidizing in limestone produce soluble sulphates of the metals and sulphuric acid, but both react readily with the limestone (calcite) giving the relatively insoluble carbonates of copper and soluble calcium sulphate; the latter is carried away. Therefore, in limestone replacement deposits the copper remained near the surface where discovery and mining were easy. The early mined deposits of the Morenci, Bisbee, and other districts were of this type.

The deposits in siliceous rocks, especially the disseminated deposits, also contained sulphides of iron and copper that, on oxidation, yielded sulphates of the metals and sulphuric acid. These did not react so quickly with the enclosing rocks, and the solutions carrying the copper moved downward until reaction with the underlying sulphide precipitated the copper from the solutions as secondary copper sulphides. Thus were formed the flat-lying disseminated deposits of copper ore, overlain by a thick mantle from which the copper has been largely leached and underlain by sulphides usually too low in copper to be ore.

There are numerous modifications to this simple relation. A change of elevation after the secondary sulphide zone has been formed may result in renewed oxidation of the secondary zone. If the earlier replacement removed most of the pyrite, the renewed oxidation may yield disseminated copper carbonate ore such as has been mined extensively in the Inspiration and Miami mines.

If the original disseminated ore is low in pyrite there may be little tendency for the copper to migrate, and oxidation may result directly in disseminated carbonate ores. This has been thought by some to explain the lack of movement of copper in the Ajo deposits where carbonate ores were practically at the surface and the grade changed but little from carbonate to sulphide ores. Many of the silver deposits of the Southwest have been mined "from the grass roots," which is to be expected in a desert region with abundant chlorides or bromides in the water which would fix the metal as the insoluble horn silver (chloride or bromide of silver).

Likewise, the oxidized gold deposits have been of the "grass roots" type with the richest ores not far below the surface. The gold has been released by the oxidation of the sulphides and in some deposits has been mechanically moved and concentrated somewhat.

This, however, does not ordinarily apply to the late Tertiary deposits like Oatman and Mogollon in which the gold is in siliceous veins with a minimum of sulphides and is but little influenced by oxidation.

From this sketchy description it is evident that oxidation has been a major factor in producing the ore bodies of the Southwest and that an understanding of the process is vital in their development.

PART II—MINING DISTRICTS

Summary of Mining History in Arizona[13]

By Robert E. Heineman[14]

Chronological Summary

1530. Nuño de Guzmán, president of the Governing Board of New Spain, hears of the Seven Cities of Cíbola far to the north, whose streets were paved with gold and silver. He founds San Miguel de Culiacán in Mexico.

1536. Alvar Núñez Cabeza de Vaca and the negro, Estevan, reach Culiacán with fresh rumors of Cíbola.

1539. Francisco Vásquez de Coronado sends Friar Marcos de Niza and Estevan to find the Seven Cities of Cíbola. Niza was the first white European to enter Arizona. The expedition reaches the Zuñi pueblos. Estevan is slain by the Indians and Niza returns with imaginative tales of gold and silver utensils.

1540. Coronado's expedition enters Arizona and discovers the Seven Cities of Cíbola to be seven Indian villages with no metallic riches.

1557. Bartolomé de Medina of Pachuca, Mexico, invents the patio process of silver amalgamation.

1583. Antonio de Espejo discovers a deposit said to be silver ore, the first in Arizona, probably the United Verde deposit, near the headwaters of the Verde River. He also discovers the Verde River salt deposits.

1604. Juan de Oñate explores northern and western Arizona but discovers no mineral.

1691. Padre Eusebio Francisco Kino enters Arizona and for twenty-five years explores and develops the Papago country.

1705. Kino mentions the mining of rich silver ores. These were probably the Santa Rita Mountain deposits.

1736. Famous Bolas de Plata silver deposit at Arizonac in northern Sonora was discovered.

1750. Some copper mined at Ajo.

1774. Placering for gold in Quijotoa district.

1776. Spanish garrison transferred from Tubac to Tucson, which up to this time was only an Indian village.

1777. Arivaca mentioned as a mining community.

1790-1820. Period of prosperity for the Spanish missions but mining was unimportant.

---

[13] Paper prepared for the regional meeting of the A.I.M.&M.E. held at Tucson, Arizona, November 1-5, 1938.

[14] Mineralogist, Arizona Bureau of Mines.

1792. Legal ratio between gold and silver in U.S. made 16 to 1.
1800. Copper being mined in a primitive way at Santa Rita, New Mexico.
1823. Mexico obtains its independence. The missions are abandoned and raided by Indians.
1825. The first American scouts begin to explore Arizona. Sylvester and John Pattee lease Chino deposit at Santa Rita, New Mexico.
1832. The scout, Pauline Weaver, first comes to Arizona.
1836. The Apaches are made unfriendly to the Americans by the murder of Chief Juan José.
1848. American troops under Kearney first enter Arizona. Marshall discovers gold in California.
1853. Gadsden Purchase from Mexico of that part of Arizona and New Mexico south of the Gila River.
1854. Charles D. Poston begins search for gold and silver near Tubac.
1855. Mexican troops leave Tucson and Tubac and are replaced by Americans. Rich copper ore hauled from Ajo to San Diego.
1856. Santa Rita silver mine near Tubac opened. Exploration of Santa Rita and Cerro Colorado Mountains.
1857. Prospectors begin to enter Arizona in numbers. Gold ore found in Mohave County near Sacramento Valley.
1858. Discovery of the Mowry lead-silver mine in the Patagonia Mountains. Discovery of the Gila City or Dome placers near Yuma.
1859. Rich silver ore being mined at Heintzelman Mine in Cerro Colorado Mountains.
1861. Start of Civil War. Withdrawal of troops followed by Apache depredations. Massacre of staff at Santa Rita silver mine with the exception of Raphael Pumpelly who escaped.
1862. Confederate troops occupy Tucson. They are driven out by California column under General A. H. Carleton who established posts at Camp Verde, Ft. McDowell, and Ft. Whipple. La Paz gold placers discovered by Pauline Weaver.
1863. Castle Dome district becomes known. Discovery of many placer and lode deposits in the Prescott region. The Moss Mine, Oatman district, Vulture Mine near Wickenburg, and Planet Mine near the Bill Williams River were discovered. The Moss and Vulture were lode gold mines, the Planet, copper. Many lode claims discovered in Mohave County. Arizona was made a territory, chiefly because of gold discoveries, with Prescott as the capital.
1864. Henry Clifton rediscovers copper in eastern Arizona.
1865. Small-scale Mexican copper operations at Cananea.
1866. Apaches on war path.
1867. Capital moved from Prescott to Tucson.

1871. The federal government sends a large number of troops and determines to end the Apache problem which was finally settled with the surrender of Geronimo in 1886.

1872. Town of Clifton founded by Metcalf and Stevens.

1873. U.S. mint by act of Congress discontinues the coinage of silver dollars. Great financial and industrial panic.

1874. Globe becomes a booming silver camp. Railroad built from Clifton to Metcalf, the first in Arizona. McCracken silver mine discovered in Mohave County. Richmond Basin district opened.

1875. Silver King Mine in Superior district discovered by Mason and Copeland. Silver Queen (Magma) also discovered. The Lesinsky brothers build a 1-ton copper furnace at Clifton. Detroit Copper Company founded and mining started at Morenci.

1876. Southern Pacific reaches Gila Bend from California. United Verde ore body discovered at Jerome by M. A. Ruffner. Mineral Park district, Mohave County, active.

1877. John Dunn, army scout, makes first location in Warren district. Ed. Schieffelin "goes to hunt for his tombstone."

1878. First shipment of matte from Copper Queen claim. First locations made at Tombstone recorded. Act of Congress again makes silver legal tender, Bland-Allison bill.

1879. Boom starts at Tombstone.

1880. Lesinskys sell out to Arizona Copper Company at Clifton after making $2,000,000. Phelps Dodge on advice of Dr. James Douglas buys half interest in Detroit Copper and builds small smelter at Morenci. Dr. Douglas pays first visit to Bisbee. Silver-copper ore is mined from Silver Queen at Superior (now the Magma).

1881. Railroad reaches Lordsburg. Old Dominion Copper and Smelting Company starts operations at Globe. Phelps Dodge acquires Atlanta claim at Bisbee. Mammoth district opened. A small copper furnace is in operation at the present site of Miami.

1882. United Verde Copper Company organized. Atlantic and Pacific Railroad crosses northern Arizona.

1883. Some copper mining undertaken at Ray. A small smelter built at Jerome.

1885. Copper Queen Consolidated Mining Company formed and builds concentrator and smelter. Territorial legislature creates University of Arizona.

1886. Bonanza ores exhausted at Morenci and concentrator built by Wm. Church to treat oxidized ore that averaged 6.5 per cent copper. Six furnaces in operation at Globe.

1887. Congress gold mine discovered by Dennis May.

1888. Dr. James Douglas turns down United Verde because of inaccessibility. Old Dominion Company reorganized at Globe. First building to house School of Mines of the University of Arizona completed at Tucson. Harqua Hala gold deposit discovered.

1889. Sen. W. A. Clark obtains control of the United Verde Mine, which resumes operations.
1890. Sherman silver purchase bill enacted by Congress. Louis D. Ricketts becomes assistant to Dr. Douglas.
1891. The cyanide process after years of experimenting becomes a success in South Africa and revolutionizes gold mining.
1892. Phelps Dodge Corporation purchases United Globe Mines at Globe and also certain claims in the Miami district.
1893. Silver demonitized. Disastrous panic. Prospectors turn from silver to gold. Copper Queen works first sulphides.
1894. Rail connection completed to Jerome. An unsuccessful attempt made to work the Ajo deposit.
1895. Cyanide process introduced at Congress Mine, one of the first installations in this country. Fortuna Mine discovered.
1896. King of Arizona Mine discovered by Chas. E. Eichelberger. McKinley elected President and gold standard assured. First disseminated copper ore treated at Clifton by James Colquhoun, but this was relatively high-grade ore.
1898. War with Spain. Captain Jim Hoatson visits Bisbee and becomes interested in the Irish Mag claim, named after one of the more famous of Bisbee's early residents.
1899. McKinley re-elected and free silver issue is dead. Daniel C. Jackling does the pioneer mill testing of a low-grade porphyry ore at Bingham, Utah. United Verde Extension Mining Company formed. An English Company, Ray Copper Mines, Ltd., unsuccessfully attempts to work the Ray deposit.
1900. A smelter is built at Douglas by the Phelps Dodge Corporation. Rich gold ore found in Oatman district. John R. Boddie, Captain Huie, and several others organize the Cornelia Copper Company to work the Ajo deposit.
1901. Railroad built to Cananea.
1902. Calumet and Arizona Company is organized.
1903. Phelps Dodge obtains control of the Old Dominion at Globe. Gold Road Mine discovered in Oatman district.
1904. Dr. F. L. Ransome of the U.S. Geol. Survey prepares report on the Bisbee district.
1905. John M. Sully examines the Chino deposit at Santa Rita, New Mexico. Waldemar Lindgren writes report on Morenci for the U.S. Geol. Survey.
1906. Philip Wiseman and Seeley Mudd obtain options at Ray. J. Parke Channing examines the deposits at Miami and exploratory shafts are started. The famous McGahan vacuum smelter is built at Ajo, the most fantastic metallurgical scheme ever devised. Copper production starts at Nacozari, Sonora. First low-grade porphyry production at Morenci when No. 6 concentrator starts to operate.
1907. Daniel C. Jackling undertakes extensive development work at Ray. John Lawler owns eight claims at Bagdad. Panic of 1907.

1908. Miami Copper Company and Inspiration Copper Company are organized. Tom Reed Gold Mines Company starts intensive operations on the Tom Reed vein in the Oatman district.

1909. Sacramento Hill at Bisbee is drilled. J. Parke Channing and Seeley Mudd drill at Ajo and reject property. Cornelia Copper Company reorganized as New Cornelia Copper Company. Louis S. Cates is placed in active charge at Ray.

1910. Hayden, Stone, & Company finance Chino, and large-scale stripping operations commence at Santa Rita, New Mexico. Magma Copper Company at Superior is formed.

1911. Production starts at Miami Copper Co. American Smelting and Refining Company builds smelter at Hayden. Ray production starts on large scale. General John C. Greenway becomes interested in Ajo and the New Cornelia property is drilled by the Calumet and Arizona Company. Production starts at Magma.

1912. Town of Oatman started. Production commences at Chino, New Mexico. Arizona admitted to the Union as forty-eighth state. James S. Douglas becomes interested in United Verde Extension and development work is begun.

1914. World War starts.

1915. Large gold ore body developed in United Eastern Mine at Oatman. Metal prices start to boom. International Smelting Company erects smelter at Miami. Flotation introduced at Inspiration, the first large-scale copper flotation plant in this country. Arizona Bureau of Mines created by State Legislature.

1916. United Verde Extension mines bonanza ore body at Jerome. A.I.M.E. regional meeting held in Arizona.

1917. United Eastern purchases Big Jim at Oatman. Production starts at New Cornelia with leaching ore. War prices for metals. Extensive working of small high-cost copper, manganese, and tungsten deposits. New Cornelia buys property of Ajo Consolidated Company.

1918. Steam shovel operations start at Sacramento Hill in Bisbee.

1919. Experimental flotation plant installed at Ajo.

1921. Postwar depression and shut down of copper properties.

1922. Entire Morenci district now controlled by Phelps Dodge. End of postwar depression.

1923. Copper Queen mill south of Bisbee is placed in operation. Rehabilitation of concentrator and new mine equipment at Chino.

1924. Ray and Chino merge. Concentrator at Ajo put into operation, and treatment of sulphide ore commences. Smelter is completed at Magma.

1925. End of high-grade ore at Miami in sight. Company plans for working low grade.

1926. Ray Consolidated absorbed by Nevada Consolidated. Large-scale leaching operations started at Inspiration.

1927. President Coolidge "does not choose to run."

1928. Drilling program started on Clay ore body at Morenci. Extensive additions to concentrator at Ajo.
1929. Climax of boom and start of the great depression. Sacramento Hill open pit operations discontinued. Louis S. Cates becomes president of Phelps Dodge. Miami Copper mining low-grade ore body successfully.
1930. Copper price collapses from 18 to under 10 cents a pound.
1931. Phelps Dodge absorbs Calumet and Arizona. Great Britain abandons gold standard.
1932. Curtailed copper operations. Extensive reworking of gold placer deposits. Four-cent tariff placed on copper imports. Copper price declines to under 5 cents a pound.
1933. Start of "New Deal." Price of gold is raised to $25.56. Silver legislation. By presidential proclamation, mint pays 64.64 cents per ounce for domestic newly mined silver.
1934. Price of gold raised to $34.95 per ounce with subsequent boom in small gold properties.
1935. Price for newly mined domestic silver raised to 77.57 cents.
1936. Period of general recovery.
1937. Business pick up, high copper prices and subsequent collapse in summer and fall. Extensive development of Clay ore body (Morenci Open Pit Mine) is started.
1938. Partial or complete shut down of copper properties and reopening in late summer. Price for newly mined domestic silver reduced to 64.64 cents. War scares. United Verde Extension finishes ore body, and smelter is dismantled and sold. Arizona Small Mine Operators Association is formed. A.I.M.E. regional meeting in Tucson in November.

PRODUCTION SUMMARY

(See graphs, Pl. IV, B.)

REFERENCES

Elsing, M. J., and Heineman, R. E. S., Arizona Metal Production, Arizona Bur. Mines Bull. 140, 1936.
Farish, T. E., History of Arizona. San Francisco, 1915.
Gardner, E. D., et al., Copper Mining in North America, U.S. Bur. of Mines Bull. 405, pp. 15-22, 1938.
Ingalls, W. R., in the Mineral Industry, pp. 225-31. New York, 1892.
Joralemon, Ira B., Romantic Copper. New York, 1934.
Lockwood, F. C., Pioneer Days in Arizona, pp. 191-217. New York, 1932.
Parsons, A. B., The Porphyry Coppers. New York, 1933.
Persons, Warren M., Barron's Annals of Business and Finance Since 1865. New York, 1937.
Quiett, G. C., Pay Dirt, 387-413. New York, 1936.
Ransome, F. L., in Bryan, Kirk, The Papago Country, Arizona, U.S. Geol. Survey Water Supply Paper 499, pp. 3-23, 1925.
Raymond, R. W., Mines and Mining West of the Rocky Mountains, pp. 321-23. Washington, 1870.

Rickard, T. A., A History of American Mining, pp. 249-300, 365-79. New York, 1932.

Tenney, J. B., History of Mining in Arizona. Arizona Bureau of Mines unpublished manuscript.

Wilson, Eldred D., et al., Arizona Lode Gold Mines and Mining, Arizona Bur. of Mines Bull. 137, pp. 16-17, 1934.

Wilson, Eldred D., Arizona Gold Placers and Placering, Arizona Bur. of Mines Bull. 142, 1937.

BISBEE DISTRICT[15]

BY CARL TRISCHKA[16]

INTRODUCTION

Each succeeding paper on the geology of the Bisbee district discloses new discoveries or modifies former ideas. This condensed account summarizes the well-established facts and presents some observations not previously recorded. A bibliography is given on page 41.

ROCKS

The columnar section (Pls. III and V) gives the essential information regarding the sedimentary rocks. It will be observed that the igneous rocks are represented as having been intruded: (1) granite in pre-Cambrian, (2) granite porphyry in pre-Cretaceous, and (3) andesite in post-Cretaceous time. This section shows also the idealized relationship of the pre-Cretaceous granite porphyry to the ore occurrences in the various formations.

GEOLOGIC HISTORY OF THE MULE MOUNTAINS

The schistosity of the pre-Cambrian Pinal schist was initiated before intrusion of the granite shown in the northwest corner of Plate VII. Accompanying this intrusion was some mineralization of which only roots of no economic importance remain.

The area was eroded to a peneplain and later sank below the sea. The Cambrian Bolsa quartzite was laid down with profound unconformity upon the schist.

The Abrigo limestone, also of Cambrian age, followed the Bolsa quartzite.

Ordovician, Silurian, and lower and middle Devonian rocks are lacking in this section of Arizona. The upper Devonian is represented by the shaly to rather pure Martin limestone. A bed of quartzite 8 to 10 feet thick is used locally as a marker between the Cambrian and upper Devonian.

From upper Devonian to early Permian time there was almost continuous deposition of limestone, followed by regional uplift,

---

[15] Paper obtained for, and originally presented at, the regional meeting of the A.I.M.&M.E. held at Tucson, Arizona, November 1-5, 1938.

[16] Geologist, Copper Queen Branch, Phelps Dodge Corporation.

erosion, intrusion of the granite porphyry, and ore deposition. The Dividend fault, with a throw of 2,000 to 5,000 feet, followed the intrusion. Erosion subsequently stripped almost all of the Paleozoic rocks from the northeast or upthrown side of this fault, exposing the Pinal schist and the intrusive granite porphyry.

An unconformity, Triassic and Jurassic in age, exists between the Carboniferous and Cretaceous. The Cretaceous consists of a great thickness of sandstone and limestone, the upper portion of which has been eroded.

During the Laramide revolution at the end of Cretaceous or early Tertiary time old fissures and faults were reopened and new ones formed. At this time weak siliceous veins were formed, and considerable tilting of the Cretaceous and other formations in the district occurred.

STRUCTURAL FEATURES

In the Bisbee Folio, Ransome recognized seven structural blocks or tracts (Pl. VI). Of these, six will be discussed briefly because of geologic interest, and the Copper Queen block will be given greater consideration because of its geologic and economic importance.

The blocks have been named: (1) Bisbee block, (2) Escabrosa block, (3) Naco block, (4) Copper Queen block, (5) Gold Hill block, (6) Glance block, and (7) Cretaceous tract.

Broadly considered, the Bisbee, Escabrosa, Naco, and Copper Queen blocks are essentially masses of pre-Cretaceous rocks, bounded wholly or in part by faults of post-Carboniferous but pre-Cretaceous age. The Escabrosa block has been dropped 2,000 to 2,500 feet with reference to the Bisbee block, whereas the Naco block has been dropped about the same amount with reference to the Escabrosa block.

The Bisbee block is composed of Paleozoic beds, pre-Paleozoic metamorphic rocks, and two intrusive bodies. The Escabrosa block has only Paleozoic rocks exposed. The limestone of the Naco Hills in the Naco block is Permian or upper Carboniferous in age.

The Copper Queen block, comprising part of a canoe-shaped syncline of Paleozoic beds, is downthrown with reference to the Bisbee block but less so than the Escabrosa block.

The Gold Hill and Glance blocks are Paleozoic and younger beds, which have been thrust toward each other from the southwest and east, respectively, over the beds of the Cretaceous Bisbee group.

The Cretaceous tract is less a definite fault block than an area of little disturbance.

COPPER QUEEN BLOCK

The Copper Queen block is bounded by the Dividend, Quarry and Bisbee West faults, and by the Black Gap fault which is east of the Campbell fault and not shown on Plate VII.

Structure.—Within this area are at least twenty fault zones, with an average strike of N. 15 to 30 degrees E. Most of them dip steeply westward, although some, particularly in the west end, dip eastward.

Crossing these fault zones are two systems of faults approximately parallel to the Dividend and Bisbee West faults. Other fracture zones either radiate from, or are concentric with, Sacramento Hill. Underground workings show radiating fault zones that contain ore where intersected by northeast fractures.

The Oliver and Dividend faults, which are cut and offset by several faults, are apparently among the oldest in the district.

The important fault zones from west to east are the Hedberg, Escacado, Czar-Shattuck, Ella, Crescent, Alhambra, Tuscarora, Dallas, Junction, Mexican Canyon, and Campbell.

Faults and fault zones, as here considered, are 100 feet in maximum width but are represented on the map by a line.

Practically all of the exposed fracture zones are traceable by the silica, manganese, iron oxides, and sparse copper glance or copper carbonate mineralization that occurs intermittently along them. Such mineralization has been followed from outcrops to underlying ore. The Glance conglomerate covers a large part of the area between the Campbell fault and the Black Gap fault to the east. Hence the pre-Cretaceous fracture zones in this area lack surface expression except where they weakly appear in the Glance conglomerate as possible upward extensions of reopened zones in the underlying Paleozoic rocks. Consequently the search for fracture zones must often be carried on underground.

Geologic history of the Copper Queen block.—After Carboniferous and before Cretaceous time erosion cut down the upper Carboniferous strata into a very irregular land surface. During this interval the area underwent doming, followed by explosive intrusion of a chimney-shaped mass of granite porphyry. The porphyry found its way also along breaks and leafed the strata apart. Probably at the same time the adjacent fault blocks settled, generally complicating the fault pattern. The main porphyry mass is surrounded by a rubble, known as the contact breccia, which consists of a mixture of all of the pre-Cretaceous rocks in the district. The contact breccia surrounds the granite porphyry of Sacramento Hill except where penetrated by dikes of the granite porphyry. The contact breccia ranges from several feet to more than 500 feet in width and is present in depth where granite porphyry masses had previously been presumed to exist.

Mineralizing solutions that came after the intrusion invaded the fracture zones in the limestone and also followed porphyry dikes and sills. Depositions of copper and iron sulphides as replacement of the limestone took place. Several waves of solution of varying mineral content, strength, and intensity must have occurred to produce the mineral zones which, though obscure, appear to be superimposed on one another. Lead and zinc sulphides are common but of very minor importance. Lead carbonate ore, however, mined in the west end of the camp, amounted to an appreciable tonnage.

The chemical composition of limestone is of slight importance compared to its physical character in determining replacement. Thoroughly broken rock along fissure zones, intersecting break zones or in embayments of the granite porphyry, was most readily replaceable. This may account to some extent for the abrupt contact between the sulphide and the limestone.

Ore replaces large and small limestone blocks and aggregates in the contact breccia.

The granite porphyry stock and the contact breccia were intensely silicified, pyritized, and sericitized. Oxidation and secondary enrichment of the mineralized granite porphyry produced a siliceous, iron-bearing gossan underlain by secondarily enriched, low-grade ore bodies.

After considerable erosion the Dividend fault, with a displacement of 2,000 to 5,000 feet, truncated the porphyry chimney. The granite porphyry of Sacramento Hill, on the south side of the Dividend fault, has gossan with low-grade ore bodies below, whereas the granite of Copper King Hill north of the Dividend fault has no gossan or ore bodies. On the north side of the fault erosion has kept pace with oxidation, and iron sulphide, with very small amounts of copper sulphide, occurs near the surface. The contact breccia around this section is composed entirely of schist. Additional evidence of the truncation by the Dividend fault is the presence of schist against the north side of the fault on the 1,800-foot level of the Sacramento shaft. Mule Pass Gulch was developed along the Dividend fault because of the weakness of the rocks.

Erosion stripped practically all of the Paleozoic rocks and any ore they may have contained from the north or upthrown side of the Dividend fault in pre-Cretaceous time.

In Cretaceous time a great thickness of sedimentary rocks was laid down, and when these rocks were tilted in late Cretaceous or early Tertiary time old breaks were opened and new ones were formed along which mineralizing solutions deposited what is known as Bisbee Queen silica. This silica replaced the basal fragments of limestone in the Glance conglomerate and also formed narrow veins in the overlying Cretaceous formations. This quartz is slightly gold bearing in places and contains sparse galena, sphalerite, and chalcopyrite, which are of no economic value. Some gulches contain small areas of gold placer from this silica.

The granite porphyry, of pre-Cretaceous age and undoubtedly of deep-seated origin, is believed to be chimneylike in the schist and to spread more widely in the fractured Paleozoic rocks.

EVIDENCE OF THE AGE OF ORE MINERALIZATION

It has been stated in several articles that because the ore mineralization in Cananea, Mexico, south of Bisbee, and in Tombstone, Arizona, is of late Cretaceous or early Tertiary age, it is reasonable to assume that the Bisbee mineralization was of that age. Proofs were offered, particularly the asserted absence of altered porphyry boulders of the Sacramento Hill type in the Cretaceous Glance conglomerate. The absence of porphyry boulders in this formation at its contact with the porphyry would argue for an intrusive contact, whereas the reverse condition would indicate an erosional contact.

The base of the Glance conglomerate, however, does contain porphyry boulders and fragments of siliceous, ferruginous gossan of the Sacramento Hill type. A short distance above this basal layer are layers of conglomerate that contain quartzite and schist. One of the best exposures showing this relation is near Sacramento Hill at the site of the old ice plant. As the Dividend fault truncates the granite porphyry but does not displace the Cretaceous, the ore must therefore be pre-Cretaceous. Underground openings on the contact of the Naco limestone (Carboniferous) and the Glance conglomerate show fragments of typical outcrop material, containing iron oxides, silica, copper carbonates, and manganese in the pre-Cretaceous detrital material which was followed to their source, an old outcrop in the Naco limestone.

Dr. Ransome's original statement that the age of mineralization was pre-Cretaceous is undoubtedly true.

SILICA OUTCROPS

Silica outcrops of two types are easily distinguished. On the surface in the western part of the district and underground in other parts of the district, the type known as silica breccia occurs. It is of pre-Cretaceous age and commonly crops out above ore. It is dense, fine grained, slightly iron stained, intensely brecciated, and cemented by silica. It occurs mostly either in the upper or in the lower Escabrosa (Mississippian) where zones of chert occur in the limestone. When the rocks were broken the chert horizons underwent intense shattering. Following the channels thus created, the mineralizing solutions not only cemented the limestone and chert fragments, making silica breccia, but in some places dissolved and removed the limestone surrounding the chert. In several places the thickness of the Escabrosa has been reduced by as much as 200 feet from its original 700 feet by this process. In several places the silica breccia itself was repeatedly shattered and recemented by silica.

The Bisbee Queen type of silica is of yellow to cream color and has an open, boxworklike structure with drusy surfaces. It occurs either on the contact between the Naco limestone and the Glance conglomerate, where it replaces the pre-Cretaceous limestone detrital material, or in veins in the Cretaceous formations. Several attempts to find ore in connection with the Bisbee Queen type of silica, which is of post-Cretaceous age, have failed. Nowhere does this silica penetrate for more than 100 feet below the old Naco limestone surface.

Ore Zones

The horizontal projection of ore bodies shows the ore zones in striking manner. As long as mining was near the Sacramento Hill porphyry stock, the zones were not clearly evident. The first to develop was along the Czar fault, a zone about 100 feet wide which contains ore intermittently along one or several breaks. Granite porphyry, as dikes and sills connected with the Sacramento Hill mass, is present in this zone.

The Southwest area ore zone is practically free of the influence of the granite porphyry. The ore is mostly in the Martin (Devonian) limestone and clusters about a siliceous core which crops out as silica breccia and manganese on the surface. This siliceous core, which is porous due to leaching, contained most of the lead carbonate ores that were mined from the district. The iron and possibly some copper were leached out of this silica core, which later collapsed. The lead, altering in place, remained in the silica. In this silica also the gold was both mechanically and chemically concentrated during leaching. Where shelves of contact material occurred between the silica and the limestone, variable thicknesses of silica resting on this contact were minable for gold.

The silica core may have resulted from complete leaching of siliceous pyrite which is commonly associated with ore bodies in all parts of the district. The oxidation slumping above the ore body as well as around its sides suggests this possibility.

The Shattuck ore zone, composed of four or five parallel fault zones, is as wide as it is long and is closely confined to the trough made by the Shattuck and Wolverine granite-porphyry dikes. This trough plunges northeast, and its bottom becomes a sill. Ore is found in and below the trough and also on the north side of the part designated as the Shattuck dike. The ore bodies were found in Shattuck and Copper Queen ground. In depth this trough becomes a dike which connects with the granite porphyry of Sacramento Hill.

The Cole-Oliver ore zone is made up of four or five parallel fault zones of considerable extent. Near Sacramento Hill the ore is associated with dikes and sills of granite porphyry. In the southern end the ore bodies are considerably elongated and dip at a low angle northward. In general the ore replaces limestone where rather strong but narrow fracture zones cut the limestone beds at right angles. The solutions evidently could not penetrate very far into the unbroken limestone and deposited their load in the narrow open channels.

The Junction-Briggs ore trend is composed of numerous fractures and fracture zones and, from the standpoint of mineralization, is next to the widest and strongest in the district. Its southern end also terminates in veinlike bodies of the type already described. In the north its ore is associated with granite porphyry.

The Campbell fault zone is being developed. At its north end the Denn Mine has found some ore bodies near the granite porphyry. Southward, mineralization that terminates in the Sorrel Horse ore-body area has been found along it.

The strongest, most important productive ore zone formed a semicircle around the granite-porphyry stock of Sacramento Hill and occurred both in the limestone and the contact breccia. This area was so thoroughly broken up prior to mineralization that the solutions deposited a major portion of their load here.

ORE BODIES

The first discovery of copper ore in the district was in the old open cut on the hillside above the Bisbee Post Office. Except at the White Tailed Deer Mine, it was the only copper ore outcrop in the district. The ore here was malachite and azurite, which for many years was the only kind of ore found or mined. As work progressed downward and southeastward, secondarily enriched sulphides and finally primary sulphides of minable grade were found.

It is clear from a horizontal projection that the ore bodies of the district are arranged in a semicircle around Sacramento Hill and also radiate outward from this center. The trend of ore which extends from northwest to southeast transverses the more steeply dipping limestones, hence the ore is found in higher formations in the southeast than in the northwestern part of the district.

Although some ore has been found in all of the Paleozoic limestones, certain beds in certain sections of the district contain more ore than those above or below them. Thus, in the Junction-Campbell area, in the eastern end of the camp, the Escabrosa (Carboniferous) limestone is favored, while in the Czar-Holbrook area on the western end of the district, the Martin (Devonian) was host to most of the ore. Between these areas, in the Gardner-Sacramento area, the Devonian and lower Escabrosa limestone are considered the most favorable rocks.

To the south, in the Don Luis area, the upper 150 feet of the Abrigo (Cambrian) is a particularly favorable formation. In the Denn Mine considerable ore was mined and discovered in the lower Naco limestone, although the major portion of it was in the Escabrosa limestone. Exploration in this area is not complete.

A commonly accepted idea about the replacement ore bodies in the limestone is that they are tabular, wider than they are high. The idea originated at a time when mining was done mostly in the western portion of the camp. Here the statement seems to apply until one considers that oxidation and erosion shrank and cut down the height of some of the ore bodies of this area. Southeastward it becomes more and more evident that the height of the important ore bodies is at least as great as their length or width, and in the extreme eastern ore area, height is generally greater than length or width. There are, of course, numerous small, rather tabular ore bodies in the district.

Oxidation progresses in intensity from southeast to northwest.

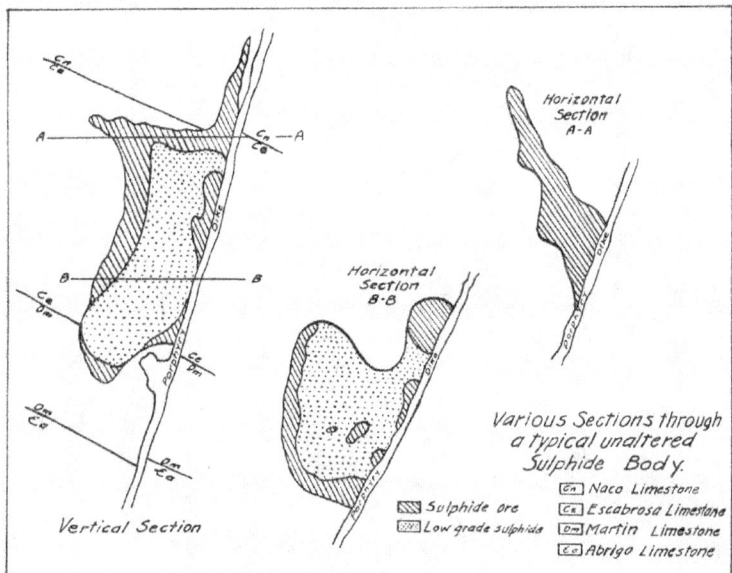

Figure 5.—Various sections through a typical unaltered sulphide ore body, Bisbee.

In small portions of the Campbell area, however, oxidation has penetrated as deep as the 2,300 level.

Ideal vertical and horizontal sections of an unaltered ore body are given in Figure 5. Practically all of the ore bodies of the district have, or had, a central core of somewhat siliceous pyrite containing small amounts of copper around which sulphides of copper and iron occurred. Oxidation and alteration obscure this relation, particularly in the northwestern part of the camp.

The pyrite core may be coarse or fine grained. In the latter case the pyrite is commonly shattered and becomes ore because of the deposition of small veinlets of copper sulphides in the breaks and cracks.

Hematite and specularite are commonly associated with the ore along its contact with the limestone. Magnetite is intimately mixed with the pyrite and chalcopyrite in certain areas.

Contact metamorphic minerals are scarce, even close to Sacramento Hill, and are practically absent at short distances from it. Characteristically there is an abrupt, clear-cut contact between the sulphide and the limestone in areas where there has been little or no oxidation. The only change discernible in most places is marbleization of the limestone. In the northwestern end of the camp where oxidation has been intense, the limestone surrounding the original ore bodies has been thoroughly saturated with the products of sulphide oxidation.

In the process of replacement the grain structure, bedding, and the included unreplaced chert lenses of the limestone are frequently beautifully preserved in the resulting sulphide.

Porphyry Ore Bodies

There is a fairly large mineralized area within the stock of Sacramento Hill. Ore in the western section of this area has been removed by steam-shovel and glory-hole mining. That in the eastern section is being mined by block caving. These ore bodies are secondarily enriched by chalcocite and are partly in the porphyry mass of Sacramento Hill and partly in the contact breccia around it. The protore contains less than 0.50 per cent copper. The stock of Sacramento Hill was highly silicified, sericitized, and pyritized, and the small amounts of chalcopyrite and bornite in the protore are responsible for the copper of the secondary enrichment. The great irregularity of the contact between the gossan and the secondarily enriched zone is worthy of note.

The porphyry on the north side of Mule Pass Gulch is not nearly so pyritized or silicified as the part in which the porphyry ore bodies occur, no secondary enrichment has taken place, and drilling found no mineralization of economic value.

Ore Guides

The ore guides in limestone may be summarized as follows:

Manganese oxides as outcrops or along fracture zones can be used as ore guides. The ore associated with them may be below or to one side of the occurrence.

Silica breccia and hematite, or both, are usually closer to ore than manganese. To get to the ore, usually found in connection with them, it is necessary to prospect down the fracture zones or the replaced bedding along which they occur.

Limonitic gossans and calcite-filled cracks in the limestone over oxidized slumped ore bodies are direct guides and point down to the possible ore.

As the calcite-filled cracks and slumping are due to either the oxidation of a sulphide ore body, pyrite body, or a solution cave, ore is not present under all of them.

Caves encountered underground are near guides, because the difference between a solution cave and a slump cave can generally be recognized.

Boxwork siderite is the result of the acid solutions which are formed when a sulphide ore body is being oxidized. The iron sulphates reacting with the limestone form siderite and gypsum. The gypsum is usually carried off in solution. The siderite forming below the sulphide which is being altered points upward as a guide. Since, however, the same solutions may come from a mass of pyrite or from a sulphide ore body, ore may or may not be present.

Granite-porphyry dikes and sills are guides to ore. By following them on both sides ore may be encountered in embayments.

Fracture zones, where they are rather steep and dip more or less normally to the bedding, are well worth following if they are at all mineralized.

BIBLIOGRAPHY OF BISBEE DISTRICT

Ransome, Frederick Leslie, Geology and Ore Deposits of the Bisbee Quadrangle, Arizona, U.S. Geol. Survey Prof. Paper No. 21, 1904.

Bisbee Folio, U.S. Geological Survey No. 112, 1904.

Douglas, James, The Copper Queen Mine, Transactions A.I.M.E., Part 2, by Arthur Notman, Vol. 29, pp. 511-46.

Torche, Wm. L., Bisbee, A Geological Sketch, M. &. S.P., Vol. 102, pp. 203-8.

Notman, Arthur, Geology of the Bisbee, Arizona, Ore Deposits, M. & E. World, March 22, 1910, p. 567.

Bonillas, Y. S., Tenney, J. B., and Feuchere, Leon, Geology of the Warren Mining District, Trans. A.I.M.E., Vol. 55, pp. 284-355.

De Kalb, Courtney, Sacramento Hill Disseminated Copper Deposits, M. & S.P., June 25, 1921.

Tenney, J. B., The Bisbee District, Fifty Years Young, E. & M.J., June 21, 1927.

Trischka, Carl, Silica Outcrops of the Bisbee District, E. & M.J., June 30, 1928.

Trischka, Carl, Rove, O. N., Barringer, D. M., Jr., Boxwork Siderite, Economic Geology, Nov., 1929.

Hewett, D. F., and Rove, O. N., Occurrences and Relations of Alabandite, Economic Geology, Vol. 25, No. 1, February, 1930.

Trischka, Carl, Bisbee Orebodies Reviewed, E. & M.J., June 8, 1931.

Tenney, J. B., The Bisbee Mining District, in Ore Deposits of the Southwest, Guide Book 14, Excursion C-1. XVI International Geological Congress, Washington, 1932.

JEROME DISTRICT[17]

BY LOUIS E. REBER, JR.[18]

LOCATION AND EXTENT

The United Verde and United Verde Extension, the chief mines of the Jerome or Verde mining district, are at Jerome, in Yavapai County, in north-central Arizona. Jerome is on the northeasterly slope of the Black Hills, facing across the broad Verde Valley to the northern Arizona plateau escarpment. The mean altitude of

---

[17] Paper prepared for, and originally presented at, the regional meeting of the A.I.M.&M.E. held at Tucson, Arizona, November 1-5, 1938.

[18] Geologist, Phelps Dodge Corp.

Jerome is about 5,200 feet, and the smelter towns of Clarkdale and Clemenceau are in the valley, about 2,000 feet lower, while the highway to Prescott climbs nearly 2,000 feet higher to cross the Black Hills. The Verde Tunnel and Smelter Railroad connects Jerome with Clarkdale which is on a branch of the Santa Fe.

The Jerome mining district includes an area of about 20 square miles, extending about 8 miles southeasterly along the steep side of the valley from near Jerome, generally between the 4,500- and 6,000-foot elevation contours. This area corresponds approximately with a belt of exposed pre-Cambrian rocks, bounded on the lower side by the great Verde fault and on the upper side by nearly flat-lying Paleozoic beds. The primary mineralization is confined to the pre-Cambrian rocks, which have been sufficiently prospected to delimit roughly the area of persistent scattered mineralization of the district and to show that it does not extend far beyond the exposed area. The southeasterly limit of the district is characterized by the gradual fading out of the scattered mineralization in the exposed belt. A short distance farther south is the Cherry Creek district of small gold veins which lie in a zone paralleling and including the contact of the Bradshaw granite.

DISTRIBUTION OF ORE MINERALIZATION

Within the district practically all the important mineralization occurs on or near boundaries of the intrusive Cleopatra quartz porphyry, two belts of which cross the district. This stronger mineralization does not appear to persist along the contacts beyond the limit of general mineralization, however.

The United Verde ore zone is on the north side of the north porphyry belt. The United Verde Extension (U.V.X.) ore zone is a downthrown fault segment of the same ore zone and shows the same relation to the porphyry. These two segments of the original United Verde ore zone account for over 99 per cent of the total production of the district. The Jerome Verde Mine is in an outlying ore body in the U.V.X. ore zone. The Verde Central ore zone is on the south side of the north porphyry belt. On the same contact, north and east of the Verde Central, the Hull-Cleopatra Dillon tunnel developed several small ore bodies, and several hundred tons were mined.

The Copper Chief-Equator (Iron King) ore zone is on the south side of the south porphyry belt. Small ore bodies, chiefly on the Green Monster property, farther north in the south porphyry belt, yielded perhaps a few hundred tons of ore. Several small veins have been worked for oxidized gold-silver ore. The Shea vein, which is near the Copper Chief, made some production, and the Grand Island Company shipped 100 tons of good ore from a small lens in the Copper Chief flat fault.

Away from the quartz-porphyry belts, two or three of the prospecting companies developed enough ore to supply the stockholders with specimens.

UNITED VERDE ORE ZONE

Location and development.—The United Verde Mine is west of Jerome. The outcrop of its ore zone was in a notch where the head of Bitter Creek cut into the relatively steep valley slope above the Verde fault (Pl. VIII). The open-pit work has removed most of the top of the ore zone, nearly to the 600-foot level. The collar of No. 3 Shaft, from which the mine levels are measured, was at an altitude of 5,500 feet. The ore zone is thoroughly developed by means of level work and diamond-drill holes to the 3,000-foot level with some work on the 3,300-foot level (altitude 2,200 feet), and a small amount of somewhat deeper diamond drilling. There is extensive stoping to the 2,550- and a little on the 2,700-foot level. Much ore remains in pillars throughout the mine.

On the 1,000-foot level the Hopewell tunnel, with a standard-gauge railroad over a mile in length, connects the mine ore bins with the surface and the railroad to the smelter. The new No. 7 Shaft, not yet in operation, comes to the surface east of the ore zone, on the 300 level bench, near the outcrop of the Verde fault (Pl. IX).

An adit tunnel on the 500-foot level connects the principal working shaft (No. 6) with the main surface plant.

There is work along the Verde fault in the Hermit claim on the 500-, 600-, 700-, and 1,000-foot levels.

Structure and extent.—The United Verde ore zone, as developed in the United Verde Mine, consists of a very irregular pipelike body of massive sulphide, quartz, and mixed sulphide and rock, with a steep north-northwesterly plunge. Quartz predominates on the hanging wall or diorite side of the main sulphide mass, with the mixed material on the footwall or quartz-porphyry side. In plan the mineralized zone ranges from more than 500,000 square feet or about 12 acres to less than 300,000 square feet, with an average near 400,000 square feet. The massive sulphide itself has an average cross section of approximately 250,000 square feet.

The downward trend of the ore zone is determined by a steeply dipping, very irregularly interfingering intrusive contact between rhyolitic quartz porphyry to the south and a series of banded tuffs and sedimentary material ("bedded sediments") to the north. It is located where the average strike of the contact changes from northerly to northeasterly. The more regularly curving contact of the diorite mass, which approximately parallels the rhyolitic porphyry-bedded-sediment contact, forms a clean-cut limit to the northerly or hanging-wall side of the ore zone. On the footwall or quartz-porphyry side the boundary is very irregular and interfingering, largely controlled by the schistosity of the porphyry, the average trend of which is about N. 10 degrees W. with steep easterly to vertical dips (Pl. XII).

In the upper part of the mine an embayment in the diorite and a band of relatively strong schistosity in the quartz porphyry combined to give the ore zone a roughly lenticular cross section, with the longer axis corresponding to the trend of the schistosity (Pl. X).

In the lower levels the more open curve of the diorite, the less intense but more uniform schistosity of the quartz porphyry, the tendency of the schistosity to approach parallelism to the contact, and, no doubt, less irregularity in the original porphyry contact, were jointly responsible for the crescent-shaped outline of the ore zone with the elongation more or less paralleling the diorite and much less interfingering with the porphyry (Pl. XI).

Although other sulphides are present, the copper content of the ore as a rule depends on the abundance of chalcopyrite. Pyrite, generally with appreciable sphalerite, constitutes the sulphide gangue. Black chlorite rock, with some quartz porphyry and quartz, is the predominant rock gangue. About one seventh of the volume of the mineralized zone is commercial copper ore, and a somewhat smaller amount is possible low-grade zinc ore.

Replacement.—As may be inferred from the preceding description of the structural features that control the form of the ore zone, the mineralization is very clearly of the replacement type. Characteristic structures and textures of the replaced rock are commonly preserved by the massive sulphide, and residual shreds of rock or unreplaced quartz phenocrysts are present in many places. Such evidence bearing on the former distribution of rock types in the ore zone aids the unraveling of the complicated history of the mineralization, which in turn serves to explain the occurrence and distribution of the commercial ore.

Sequence of mineralization.—Several definite stages of deposition can be recognized, although the extent to which they represent distinct periods in the sequence, rather than parts of a more or less continuously progressive change, is not clear.

Paragenesis or sequence of mineral deposition, as conventionally determined by the microscopic study of polished surfaces, may prove misleading unless interpreted in the light of detailed study of the occurrence and distribution of the material in place. The correct understanding of the most significant features of a complicated chronology may depend much more upon field study than upon the microscope. Nevertheless the results of microscope work are also essential to a complete understanding.

Microscope work by Benedict,[19] Lindgren,[20] and Hansen,[21] supplies such data.

Two points not made clear by the microscope are significant. First, although breccia filling and replacement of earlier by later sulphides were quantitatively important, rock replacement was most important from beginning to end, so that in a broad way the distribution and structure of any generation of sulphides was

[19] P. C. Benedict, *Geology of Deception Gulch and the Verde Central Mine,* unpublished thesis, Mass. Inst. of Technology, 1923.

[20] Waldemar Lindgren (U.S.G.S. Bull. 782, 1926).

[21] M. G. Hansen, *Microscopical Examination of the United Verde Sulphide Orebody,* unpublished report to United Verde Copper Company, December, 1927.

primarily controlled by the location of the most accessible unreplaced rock left by preceding generations. Second, the bulk of
the black chlorite was formed after deposition of most of the lean
pyrite and before most of the chalcopyrite.

The most plausible interpretation of the mineralization is then
as follows: The first solutions followed sections of the porphyry
contact nearest the diorite and deposited quartz with negligible
quantities of pyrite and chalcopyrite and probably specularite.
The quartz favored replacement of the bedded sediments but in
places left a narrow unreplaced strip against the diorite.

The second period of mineralization deposited pyrite with important quantities of dark red-brown sphalerite or marmatite and
a little chalcopyrite, with probably some local quartz as well as
intergrown quartz and dolomite similar to that of the sixth period.
Microscopic arsenopyrite preceded the pyrite. This period is
responsible for the major part of the ore zone, and probably a
large part of the possible zinc ore, but no commercial copper ore.
Some bedded sediments were replaced, but the pyrite appears to
have shown a preference to replacement of the porphyry. For the
above reason, or because it was already sealed off by the quartz,
the strip of bedded sediments along the diorite remained unreplaced. Microscopic magnetite and a very small amount of specularite following the pyrite were perhaps precursors of the change
from sulphide to high-iron chlorite deposition.

The third period represents a definite break in sulphide deposition. Solutions working out from the footwall of the pyrite sulphide mass completely replaced an enormous volume of quartz
porphyry and some tongues of the bedded sediments with a
nearly black, high-iron variety of chlorite.

In the fourth period were deposited much pure chalcopyrite
and considerable chalcopyrite intergrown with nearly black
sphalerite or marmatite and in places with galena and probably
some pyrite. The chalcopyrite appears to have most readily replaced the black chlorite rock or "black schist." This period was
responsible for most of the commercial ore. Numerous tongues
and lenses of "black schist" interfingering with the earlier sulphide were completely replaced, material additions were made to
the total volume of massive sulphide, and some scattering mineralization formed ore bodies in the schist but very rarely penetrated the quartz porphyry.

The structure that controlled the form of the ore zone in the
upper and lower levels is also significant. In the upper levels
there was evidently much more interfingering of replaced and
unreplaced rock before the chalcopyrite came in, and by further
replacement it penetrated deep into the earlier lean sulphides.
In the lower levels the earlier sulphide mass was comparatively
tight, and for the most part the chalcopyrite was forced to build
onto the footwall of the pyrite or spread out into the schist.
Probably the greater steepness and regularity of the diorite wall
in the lower levels also militated against a more favorable distribution of the early sulphide.

After the fourth-period mineralization numerous small andesite or fine-grained diorite dikes cut through the ore zone and surrounding rocks along nearly vertical east-west fractures at intervals of 100 to 200 feet or less. These dikes range in thickness from a few inches to 20 feet, although most of them are less than 2½ feet. Such dikes are common throughout the district but nowhere else so abundant. Though somewhat mineralized locally, characteristic differences in the fracturing where they cut high-grade chalcopyrite from that in the lean pyrite are believed to prove them younger than the principal chalcopyrite stage of deposition.

During the fifth period small masses of quartz, in part associated with considerable bornite and probably some other sulphides, were deposited. Microscopic primary chalcocite is intergrown with bornite, and recrystallized black chlorite is an occasional associate.

The sixth and last period was characterized by widespread deposition of intergrown quartz and dolomite or calcite in part associated with chalcopyrite; pyrite; a relatively clear, glassy, pale brown or greenish variety of sphalerite; and tennantite. This mineralization is most conspicuous in the chlorite rock or black schist and probably added materially to the volume of schist ore. The tennantite, predominantly arsenical, contains some antimony and about 40 ounces of silver to the ton, but it rarely affects the silver content of the ore. The same mineral association deposited in fractures in the older quartz and sulphides and in small gash veins in the schist, in places with margins of fibrous serpentine, is the final phase.

Changes with depth.—The ore zone in the U.V.X. Mine probably represents a segment from over 2,000 feet above the top of that exposed in the United Verde Mine. Probably a large part of the chalcocite ore was of fairly good grade before enrichment. As in the highest levels in the United Verde, there was probably a smaller-than-average area of mineralization, with a higher-than-average proportion of chalcopyrite.

In the United Verde Mine the size of the ore zone and the amount of copper ore both vary greatly from level to level. Both increase where the plunge of the ore body is less steep than average.

In a very broad way, the trend is a diminishing one from the top to the bottom of the mine as regards quantity of ore, and from about the halfway point to the bottom as regards the size of the ore zone. The ore zone at the bottom is somewhat larger than the minimum between the halfway point and the surface.

The structural features are no doubt to some extent responsible for the diminishing quantity of ore.

If the lower-level trend of the ore zone and the dip of the east end of the diorite persist, the ore zone may leave the diorite altogether, in which case some scattering of the ore zone, with perhaps discontinuous lenses of massive sulphide, would be expected.

This change would probably mean less commercial ore in proportion to the total copper mineralization but might have some favorable aspects. There is a chance for the ore zone to flatten out before it entirely leaves the diorite.

A mineralized zone, with an area of about 10,000 square feet of massive sulphide, quartz, and mineralized schist, encountered on the 3,000-foot level, west of the diorite on the U.V.X., Haynes property, may represent the top of a branch from the main ore zone. If so, the ore zone may become stronger below the junction.

The Haynes-area mineralization includes considerable magnetite and a small amount of pyrrhotite, although otherwise typical of the main ore zone. This may be the precursor of an unfavorable mineralogical change with depth. The mineralization has been remarkably constant for a vertical range of over a mile, from the highest primary mineralization in the U.V.X. Mine to the bottom of the United Verde. Microscopic pyrrhotite in the main ore zone on the 3,000-foot level and more microscopic magnetite and specularite than above, tend to confirm the Haynes showing.

Analyses.—The analyses in Table 3, though not exact averages, give the approximate chemical composition of the material of the primary mineralized area. The last two show the change effected in replacement of quartz porphyry by chlorite.

In addition to the material represented by the analyses, a considerable volume of lean siliceous sulphide and quartz and a smaller amount of mineralized bedded sediments make up the volume of the mineralized zone.

Abundant black chlorite in the ore added greatly to smelting difficulties. In 1927 the concentrating plant was added to the smelter at Clarkdale primarily to eliminate excess black schist from the smelting charge rather than to treat lower grade ore.

Sulphide enrichment.—Although apparently unenriched massive pyrite was in places close under the oxides, it is believed that chalcocite enrichment appreciably affected the chalcopyrite ore as deep as the 500-foot level. Possibly a considerable part of the highest grade ore on the 300-foot level was chalcocite. Detailed records of the early mining are lacking, and pit operations encountered most of the highest grade pillar material as crushed and broken fragments often mixed with old stope fill. Much chalcocite and considerable bornite were present in the most crushed material from the pillars under the 300-, and less extensively under the 400-foot levels. Conditions did not permit even an approximate estimate of the total amount or the mode of occurrence of the chalcocite and bornite encountered in the pit.

Bornite and steely chalcocite were found as lumps and boulders in loose, porous material showing evidence of intense fire action. The time from the first mine fires to the opening of the pit was about thirty years, during much of which the material was extremely hot. The appearance and occurrence of the bornite sug-

TABLE 3.—AVERAGE ANALYSES (PER CENT).

	Cu	Zn	SiO$_2$	Fe	Al$_2$O$_3$	S	CaO	MgO	K$_2$O	Na$_2$O	H$_2$O	CO$_2$
Lean massive sulphide	0.70	1.33	11.00	35.00	1.39	37.00	2.07	......	......	......	......	......
Zincky massive sulphide	1.16	6.72	11.50	33.00	1.18	38.00	0.78	......	......	......	......	......
Massive sulph. copper ore	4.99	2.40	11.80	31.40	3.50	32.60	1.80	......	......	......	......	......
Siliceous sulph. copper ore	5.44	1.90	37.20	21.40	3.70	21.30	0.64	......	......	......	......	......
Schist copper ore	4.73	0.70	22.90	20.70	12.70	15.40	1.22	......	......	......	......	......
Quartz porphyry copper ore	2.79	0.90	40.10	15.70	12.50	10.80	1.05	......	......	......	......	......
Quartz porphyry	0.02	0.04	72.08	2.21	14.57	0.06	1.98	1.48	2.38	1.26	2.67	1.35
Black schist (chlorite)	0.03	0.26	26.23	15.59	25.99	0.11	0.11	18.16	......	......	8.55	0.04

gested that it might have formed in the fire area. Boyd[22] in 1935 concluded that all or nearly all of the bornite and probably much of the chalcocite were formed by fire action. He produced bornite by maintaining chalcopyrite at a temperature of 500 degrees Centigrade in a reducing atmosphere for four to five hours. With further heating some bornite changed to chalcocite. It is thus likely that much of the chalcocite in the pit was not due to original secondary enrichment. Furthermore, the absence in the oxide zone of any such accumulation of quartz as characterized the gossan over the U.V.X. chalcocite, makes it probable that the former chalcocite zone over the United Verde was much weaker than that of the U.V.X., and that enrichment before the pre-Cambrian faulting accounts for this difference.

Sooty chalcocite and covellite, sufficient to affect ore values, were found in crushed material below the 500-foot level. This enrichment was of comparatively recent date and presumably due largely to fire-zone conditions.

A body of disseminated ore in the quartz porphyry adjoining the main ore zone, about 500 feet in vertical extent and containing about 1,000,000 tons of 1.5 per-cent ore, was formed by secondary enrichment of very lean disseminated pyrite. The very small amount of leached capping and the prevalence of traces of oxidized copper mineral throughout indicate this enrichment to be not related to the present erosion cycle.

The distribution of the gold and silver in the open pit and comparison with primary ore at greater depth led to the conclusion that there has been secondary enrichment of both gold and silver. Possibly the precious-metal enrichment also had some relation to artificial stimulation by fire-zone conditions.

Oxide zone.—The gossan of the United Verde ore zone occurred as a blanket about 100 feet in average thickness with the low point slightly below the 160-foot level of the mine. It consisted largely of rather highly colored, soft limonitic material with lenses and boulders of hard iron oxide. At the south end copper carbonate minerals were conspicuous where mineralized black schist interfingers with the quartz porphyry. On the north end massive primary quartz cropped out against the diorite. Lower and to the south of the massive quartz and to some extent along the east side of the gossan, were a few prominent exposures of brecciated, honeycombed quartz. The greater part of the soft gossan was no doubt covered with soil and boulders of the hard limonite.

High-grade gold-silver ore was mined from parts of the soft gossan, and several places showed high concentrations of native silver immediately overlying the sulphides. When mined from the open pit, almost all of the soft gossan proved to be of commercial grade. Portions with relatively low gold-silver were generally high enough in silica to make converter flux.

---

[22] L. H. Boyd, *Microscopic Examination of Certain Ores from the United Verde Fire Stopes,* unpublished thesis, Colorado School of Mines, December, 1935.

The following analysis (in per cent) is of oxide ore shipped during 1918:

Cu	Fe	Zn	SiO₂	Al₂O₃	S
1.42	31.5	0.2	34.1	5.7	4.2

The Hermit claim has about 230,000 tons of oxidized copper ore, chiefly azurite and malachite with some chrysocolla, copper pitch, and black copper oxide, in basalt and prebasalt gravel in the hanging wall of the fault. The copper was evidently transported from the top of the United Verde ore zone.

U.V.X. ORE ZONE

Location and development.—The U.V.X. Mine is below the fault on Bitter Creek where it parallels the lower side of Jerome (Pl. VIII). The collar altitude of the original Daisy shaft is 5,050 feet, and the bottom of the mine is the 1,900-foot level at 3,200 feet altitude. Some diamond-drill work has been done in the ore zone below the 1,900-foot level, and the outlying area, formerly Jerome Verde property, is explored most extensively on the 1,400- and 1,700-foot levels. The Josephine tunnel, with a standard-gauge railroad 2¼ miles long, connects the shaft ore bins with the surface.

Structure and extent.—The Daisy shaft passed through the fault into the pre-Cambrian rocks at 4,600 feet altitude, whereas the main shafts are in younger rocks to about 4,300 feet altitude.

The top of the ore zone ranges from a little above the 700-foot level to a little below the 800 level, with 500 to 700 feet of overlying younger rocks. Although a wedge of Paleozoic formations overlies part of the ore zone, most of it was exposed by a deep Tertiary canyon so that prebasalt gravel rests directly on the gossan (Pl. XIV).

The ore zone on the 800-foot level, near the top of the pre-Cambrian (Pl. IX), is of about the same size or a little larger than the top of the United Verde ore zone.

The form and trend of the U.V.X. ore zone has been controlled primarily by the greenstone-quartz porphyry contact and to a less degree by the margin of the diorite. The general east-west trend of the interfingering porphyry contact and the local schistosity parallel to that trend account for the direction of the long axis of the main ore body.

As shown on the 1,300-foot-level map (Pl. XIII), the main ore body, with maximum east-west length of about 500 feet and width of about 200 feet, constitutes the most important part of the ore zone. From the main ore body bands of mineralization extend northwesterly around the northeast side of the diorite and penetrate re-entrants in the diorite. In the upper part of the ore zone the northeast diorite contact dips northeast and the southeast contact dips generally southeast; the re-entrants are to some extent southeasterly plunging troughs in the diorite. Small veinlike tongues with a northwesterly plunge extend south and east of the main ore body along shear zones related to the interfingering of porphyry and greenstone.

In the lower part of the ore zone, the southeast or south side of the diorite has a northerly dip and the east part a westerly dip that corresponds with dips in the United Verde. The ore zone which consists of lenses of massive sulphide joined by weaker mineralization plunges northward. The pinching out of the main lens of massive sulphides below the 1,600-foot level, at an altitude of about 3,450 feet, forms a sag in the footwall of the ore zone entirely comparable to the footwall structure in the United Verde Mine. The deeper sulphide lenses are generally closer to the diorite, though also related to the porphyry-greenstone contact. Unlike the United Verde Mine, the U.V.X. Mine contains much fine-grained, schistose marginal diorite, some of which has been mineralized. The apparent penetration of the diorite by massive sulphide, however, is partly due to faulting.

Faulting.—From near the pre-Cambrian surface to below the 1,900-foot level the U.V.X. ore zone is progressively cut off by the Verde fault.

The possibility of pre-Cambrian postmineral faulting was always given due consideration but not regarded seriously until work in the U.V.X. Mine proved the complete cutoff of the ore zone by the fault and began to suggest that there was no footwall continuation. The predominant trend of striations and rolls in the fault indicates a direction of movement not more than 10 to 20 degrees southeast of the dip. This direction, with the measurable vertical component, pointed to the position of the footwall segment, although the evidence was not deemed conclusive enough to justify definitely limiting prospecting along the fault. Exploratory work by 1926 made it fairly certain that there was no segment of the ore zone under the fault and established the corollary of pre-Cambrian displacement. By projecting all geologic data on fault-plane sections, the writer showed conclusively in 1928 that the opposite sides of the fault could not be made to match by reversing the Tertiary movement, regardless of assumptions as to lateral movement. This fact established the former continuity of the United Verde and U.V.X. ore deposits beyond reasonable doubt. Projecting the United Verde ore zone and the plane of the fault to an intersection above the outcrop gives a vertical displacement of about 4,000 feet, or possibly anywhere between 3,500 and 4,500 feet. Projecting the base of the Paleozoic rocks to the plane of the fault from opposite sides gives 1,600 feet for the Tertiary vertical displacement, which is probably exact within 50 feet. Hence the pre-Cambrian displacement was 2,400 feet plus or minus 500 feet. These figures check with Ransome's interpretation.[23]

In general nearly 90 per cent of the Tertiary Verde fault movement is connected along the Verde fault, a single plane with heavy gouge sometimes referred to as the main footwall break. The remainder of the Tertiary and possibly some of the pre-Cam-

[23] F. L. Ransome, *Ore Deposits of the Southwest* (16th Int. Geol. Cong., 1932), Guidebook 14, pp. 20-21 and Pl. 4.

brian movement is distributed over a braided system of small hanging-wall breaks that are extremely difficult to map accurately and may have considerably affected the structural details of the sulphide masses.

Due to several much stronger hanging-wall branches farther north and up the dip of the fault, the footwall strand under the small sulphide body mined from the United Verde 500-foot level represents only about half of the total Tertiary displacement.

Sulphide zone.—The top of the sulphides of the main ore body is about 50 feet below the 1,200-foot level. Over the most important vein ore body, it is near the 1,100-foot level. Small sulphide masses in the quartz and some sulphide in the southeast veins range from altitudes of 4,300 to 4,400 feet, with a few fairly close to the pre-Cambrian surface.

Although primary quartz and residual pyrite are present, the primary mineralization is masked by intense secondary enrichment throughout much of the mine. There was enough primary material in the deeper parts of the mine, however, to permit comparison with the United Verde deposit.

The hard, dense, lean pyrite with microscopic quartz, which is conspicuous in the United Verde ore zone, is scarce in the U.V.X. ore zone, which contains a greater abundance of quartz-carbonate gangue. Most of the minerals of the United Verde deposit are present in the U.V.X., and all of the primary material in the U. V.X. deposit can be duplicated in the United Verde deposit. The black chlorite rock or "black schist," with the characteristic interfingering pattern, is extensively developed by replacement of quartz porphyry and siliceous greenstone in the vicinity of the main U.V.X. ore body. Farther north it replaces less siliceous greenstone and minor quantities of schistose diorite as well as quartz porphyry.

The sequence of mineralization appears to have been similar to that in the United Verde Mine.

The typical lean primary sulphide in the U.V.X. Mine consisted of pyrite with minor quantities of sphalerite and chalcopyrite (2 to 4 per cent zinc and ½ to 1½ per cent copper), in places banded, with admixture of quartz carbonate, and cut by numerous quartz-carbonate veinlets.

The lower part of at least one of the smaller ore bodies was very good, clean, primary chalcopyrite ore. Some and perhaps much of the slightly enriched ore owed its value to primary chalcopyrite. The proportion of chalcopyrite in the primary material does not differ greatly from that in the United Verde Mine.

The study of polished sections by Lindgren[24] and Schwartz[25] indicates the quantitative importance of pyrite replacement, and

[24] Waldemar Lindgren, *Ore Deposits of the Jerome and Bradshaw Mountains Quadrangles, Arizona* (U.S.G.S. Bull. 782, 1926), pp. 54-97.

[25] G. M. Schwartz, "Oxidized Copper Ores of U.V.X. Mine," *Econ. Geol.,* XXXIII (January, 1938), pp. 21-33.

Lindgren[26] has expressed the view that there never was much chalcopyrite in the ore. The writer believes that chalcopyrite replacement was most important in portions of the high-grade chalcocite ore. If this is true, proportion of primary chalcopyrite in the U.V.X. deposit corresponds to a better-than-average section of the United Verde deposit.

Sulphide enrichment.—The intensity and extent of the secondary enrichment in the U.V.X. Mine formed an almost unique deposit of chalcocite that places the mine in the front rank of high-grade copper mines. Outside of the main ore body the principal lenticular veinlike body was a small bonanza in itself, and numerous smaller bodies helped to swell the high-grade total.

The decrease of chalcocite with depth, the general scarcity of chalcopyrite or sphalerite where chalcocite was most abundant, and the intense kaolinic alteration of the wall rocks, varying with the abundance of chalcocite, conclusively indicate formation by the process of secondary enrichment, which is also confirmed by microscopic evidence.

The evidence only shows that all but a very little of the chalcocite was formed before the deposit was covered by the Paleozoic sediments. Since then, except for a fleeting instant in geologic time, after the precipitous prebasalt canyon barely cut into the top of the gossan, the deposit has been continuously buried. A minute quantity of sooty chalcocite and covellite has been formed by underground-water circulation since the faulting. The basal Tapeats sandstone is not believed to be older than middle Cambrian, and important secondary enrichment could have taken place in early Cambrian time. The same reasoning applies to the date of the earlier postmineral faulting; but evidence from comparison with the United Verde deposit indicates that the greater part of the U.V.X. enrichment preceded that faulting; and minor deformation, thrust faulting, and probable interrelation of the intrusive rocks point to a not very late pre-Cambrian age for the primary mineralization and make it most probable that the age of the enrichment was truly pre-Cambrian.

Oxide zone.—The bottom limit of the oxide zone in the U.V.X. Mine is extremely variable, ranging from an altitude of about 3,850 feet over the main ore body to more than 4,400 feet at the tops of some of the smaller sulphide bodies. Above the 1,200-foot level, however, dense fine-grained primary quartz, much of which carries very little sulphide, becomes more and more predominant. Within the quartz were a number of irregular lenses of high-grade quartz-chalcocite ore, some of very limited vertical extent. In the higher bodies malachite, locally with a little cuprite, was conspicuous but generally unimportant.

The gossan or capping over the main ore body is extremely siliceous and includes much massive quartz which in part shows repeated brecciation and recementation, but in part is difficult to distinguish from the undisturbed primary quartz. Somewhat

[26] Waldemar Lindgren, unpublished report of U.V.X. Mine, August, 1926.

cavernous, hard quartz breccia with considerable limonitic material is more abundant than the cleaner quartz. Soft limonitic material, comparable to most of the United Verde gossan, occurs only very locally, although predominant close to the top of the sulphides, with much native copper in places. The vertical extent of the gossan ranges from 450 to 500 feet. Some of the smaller quartzy ore bodies were capped with 40 to 50 feet of relatively irony gossan, but in a few places the chalcocite merged into massive quartz with no obvious leaching or slumping.

The veinlike tongues extending south and east from the main ore bodies, not terminated by weaker mineralization, were overlain by 50 to 100 feet of thoroughly leached, kaolinized rock with more or less limonitic material. The vertical range of the ore bodies was from near the 1,300-foot level to a few floors above the 800-foot level. Oxidized copper minerals which were prevalent throughout, in some of the higher parts accounted for over half the copper content and entirely masked the finely divided chalcocite. All the more common oxidized copper minerals were present, with malachite, chrysocolla, and azurite most abundant. In the deeper parts sulphides with some small stringers and lenses of massive chalcocite and bornite were generally more conspicuous.

Such partially oxidized copper ore yielded nearly one eighth of the production of the mine, whereas ore with all the copper in oxide minerals, mined from the prebasalt gravel or conglomerate in the northwest part of the mine amounted to about 2 per cent. The conglomerate ore was formed at a comparatively recent date by ground water carrying copper presumably from the top of the United Verde ore zone and is similar to ore developed by the United Verde along the fault.

Unlike the United Verde gossan, even the soft limonitic material was rarely commercial gold-silver ore in the U.V.X., although some high-grade native silver ore occurred above the chalcocite of the main ore body.

The "gold stope" ore body was a tabular veinlike body along the diorite contact, bottoming in a trough in the diorite, and in part limited by massive quartz. The typical ore was fine-grained friable quartz sand with almost no residual iron oxide. The maximum length was about 350 feet, the width from 5 to 20 feet, and the vertical extent nearly 200 feet. It extended above and below the 950 level from an elevation of about 4,060 feet to about 4,250 feet. It may have averaged $10 per ton in gold, with some relatively high-grade sections. Evidently, the local conditions were exceptionally favorable to concentration of gold.

COPPER CHIEF-IRON KING ORE ZONE

Location and development.—The Copper Chief and Iron King or Equator mines are 3½ miles south-southeast from Jerome (Pl. VIII). The two mines are separated by a property boundary that divides the ore zone, with the Copper Chief to the west and the Iron King to the east. The collar of the Copper Chief shaft was within the principal gossan area at an altitude of about 5,750 feet. On the bottom level, about 350 feet lower, an adit tunnel connects the shaft with the surface 850 feet to the south. There was fairly extensive stoping of oxide ore from the 240-foot level to the surface on the Copper Chief side of the property line. Mill holing of ore and waste fill has formed a considerable pit at the surface corresponding to the original extent of the gossan. All the Copper Chief work above the 240-foot level is inaccessible. On the 300-foot level are some stopes in massive sulphide.

The Iron King Mine was worked through an adit from the east about 80 feet above the Copper Chief adit, with a length of about 500 feet from the portal to the property line. Stoping extends from about 15 feet below the tunnel to about 70 feet above, with several raises, winzes, and sublevels.

Structure and extent.—The Copper Chief-Iron King ore zone extends for about 800 feet along a vertical premineral fault contact that trends about N. 80 degrees E. between quartz porphyry on the north and banded greenstone tuff on the south. It has been formed very largely by replacement of the greenstone; the quartz porphyry makes a fairly regular, nearly vertical north wall. West of the property line the deposit has the form of the west half of a fairly regular lens, with the south side somewhat ragged due to the influence of the low-angle northerly dip of the greenstone banding. The surface gossan was about 60 feet wide near the property line and extended about 250 feet west. The deposit has a somewhat smaller horizontal area at the top of the massive sulphide, about 240 feet down, and bottoms on the banding of the greenstone at about 320 feet.

Near the property line the top of the gossan plunges to the east, leaving on the surface only a narrow streak that follows the contact down to a mass of gossan at the portal of the Iron King tunnel and a few less-definite streaks along the trend of the south side. In the Iron King end the south or footwall side of the ore zone is more completely controlled by the rock banding so that vertical sections tend to have an unsymmetrical inverted crescent form, with the bottom progressively thinner eastward until it pinches out.

At or near the bottom the porphyry contact leaves the steep fault and continues with a relatively flat north or northwesterly dip. No downward continuation of the mineralization has been found other than a very small chimney or pluglike mass of lean sulphide, about 600 square feet in area, which trends vertically downward in the greenstone.

Copper Chief fault.—The Copper Chief fault, which has a very flat west to northwesterly dip, cuts through the Iron King tunnel near the portal, passes under the ore zone, and shows on the Copper Chief tunnel level. It can be traced on the surface for about ¼ mile to the north and ¾ mile to the southwest from the Iron King portal (Pl. VIII). A rather extensive diamond drilling campaign was based on the possibility of the fault having cut the ore zone and someone's opinion that the fault represented a large overthrust from the northwest. At a later date the Iron King winzes were unwatered and the small chimney of massive sulphide rediscovered on the 100-foot sublevel. A small amount of work to improve the exposure of the fault brought to light evidence of grooving and drag and gave a nearly exact measure of the amount and direction of the fault displacement. This indicated a horizontal throw of a little over 300 feet in a direction more nearly east than southeast. The negative result of the exploratory work indicated by this determination made it fairly certain that there was no displacement of the bottom of the main sulphide lens.

Sulphide zone.—Although there is much less fine-grained jaspery quartz, and black chlorite is conspicuously absent; the general character of the sulphide mineralization has enough in common with that in the United Verde to indicate a close relationship. Any type of sulphide aggregate in the Copper Chief-Iron King ore zone is duplicated somewhere in the United Verde.

Sulphide enrichment was limited on the Copper Chief side to a 2- to 4-foot layer where sooty chalcocite and covellite merged downward into crumbly pyrite and upward into oxidized material, in places with some concentration of copper and lead carbonate, and to a similar but generally thinner layer on the Iron King side. Some of this material was high in silver.

Probably about half of the sulphide in the ore zone was ore. Nearly all of it was in the Iron King end, from which about 30,000 tons were smelted. The Copper Chief shipped a few hundred tons of sulphide ore when there was a good copper market, and more recently lessees have shipped about 6,000 tons of low-grade sulphide to the U.V.X. company for flux.

Oxide zone.—The oxide ore mined by the Copper Chief and Equator companies was very similar to the soft limonitic gossan of the United Verde, although the average lead content was no doubt somewhat higher. The few thousand tons of lower grade oxide left in the Iron King by the Equator Company and now being taken out by a lessee are decidedly less irony and considerably higher in lead.

Verde Central Ore Zone

Location and development.—The Verde Central Mine is south of Walnut Gulch near the Jerome-Prescott highway and about 4,000 feet south of the United Verde Mine.

The collar of the main shaft is at an altitude of about 5,500 feet. The bottom level is the 1,900 foot at an altitude of 3,594 feet. In addition to the development of the productive area, there is extensive exploratory work to the south on both the 100- and 1,000-foot levels and a long crosscut to the west on the 1,450-foot level.

Structure and extent.—The Verde Central ore zone, like the United Verde and U.V.X. zones, is where the contact of the quartz porphyry is extraordinarily irregular or interfingering, although on the footwall instead of the hanging wall of the porphyry mass. The exceptionally long tongue of porphyry extending south from above the Verde Central and the change in the average trend of the contact are probably most significant. The Verde Central was a "blind" prospect because the plunge of the interfingering contact zone brought the mineralization into the workings below the surface. The surface showing of irony gossan, altered black schist, and quartz with some oxidized copper minerals, around the end of the porphyry tongue on top of the hill to the south at an altitude of about 5,850 feet, undoubtedly is the outcrop of the Verde Central ore zone (Pl. IX).

There are two types of ore in the mine. One on the contact extending south from near the shaft is patchy mineralization in black schist. This has produced an ore body which extended from above the 800- to below the 1,000-foot level, with a length of about 200 feet and a mean width of about 9 feet.

The other type, the one responsible for most of the Verde Central ore, consists of a tabular, veinlike body of quartz, pyrite, and chalcopyrite in varying proportions, which though irregular in detail and varying in width from 5 to 30 or 40 feet, is essentially continuous for a maximum length of over 1,000 feet on the 1,000-foot level. It begins to show near the top of the mine, and is still fairly strong at the bottom. Although there is much less chalcopyrite than quartz or pyrite, its distribution is such that bodies of good size and an average copper content of nearly 3 per cent can be mined. Below the 600-foot level, ore of this character made up a fair proportion of the vein. Mining more selectively to produce higher grade ore would be very difficult, however.

It required abnormally high copper prices to enable the Verde Central to enter production, and the net profit on the copper produced plus any additional profit that may result from the developed ore remaining in the mine, must fall far short of returning the cost of the exploratory work.

Sequence of mineralization—Three stages of deposition—quartz, pyrite, and chalcopyrite—or four, if the black chlorite be included, correspond to the principal periods of deposition in the United Verde deposit. A vertical channel pattern of the ore shoots in the vein points to a definite separation between the pyrite and the chalcopyrite periods, although there is not much indication of the black chlorite solutions having followed the main pyrite channels as might be expected from comparison with the United Verde. For the most part, the black chlorite appears to have worked out from the porphyry contacts. What appears to be an abnormally low precious-metal content may be a function of the scarcity or absence of the other sulphides common as minor constituents of the United Verde deposit.

Supergene, secondary sulphides and oxide zone minerals are of negligible interest or importance in the Verde Central ore bodies, although a few burro loads of carbonate ore and chalcocite were found associated with the outcrop of the ore zone.

SHEA ORE ZONE

Location and development.—The Shea Mine is about 1,500 feet south of the Copper Chief outcrop. The shaft follows a south-dipping vein for 1,220 feet at an average dip of 42 degrees, attaining a vertical depth of 825 feet. The lower tunnel level, which reaches the surface about 1,300 feet east of the shaft, has almost 3,000 feet of drifts. The total of all the level work is about 7,000 feet.

Structure and extent.—Though replacement was no doubt a factor, the Shea vein is a clean-cut quartz vein that follows a premineral fault of a probable throw of more than 100 feet. The vein can be traced for about 1,200 feet east from the portal of the main adit, making the known length over 3,400 feet. The most westerly 650 feet of the adit level drift is on a different fracture. The surface exposure ends less than 200 feet west of the shaft where blanketed by the flat Copper Chief overthrust fault. Farther west the vein is weak immediately under the fault, and the fissure has not been positively identified on the surface west of the flat fault outcrop. In the most westerly part of the mine and from some distance west of the tunnel portal to the east limit of exposure, the quartz is lenticular and discontinuous. It may not average more than a foot in thickness throughout the mine, although the maximum in the vicinity of the ore shoot exceeds 5 feet.

The vein is for the most part within a large area of the dark, moderately fine-grained dioritic rock which has been called the Shea diabase. Near the shaft, it cuts across the north-south granite-porphyry dike belt.

The flat fault has been explored from the mine workings near the surface and drifted on for short distances from two deeper points, one about 500 and one about 1,200 feet west of the outcrop. This latter indicates an average dip of only 13 degrees for the flat fault.

Mineralization.—A large part of the vein consists of coarsely crystalline white quartz with sparsely scattered pyrite and only a few spots or bunches of very coarse intergrown ankerite or siderite and pyrite, usually with some chalcopyrite. Locally, there is also a little arsenopyrite and some tetrahedrite. The tetrahedrite (or freibergite) carries 600 ounces or more of silver per ton. It contains only a little arsenic. Where the sulphide mineralization is strongest and the tetrahedrite most abundant, the vein has a banded structure due to variation in the abundance of the minerals. Near the ore shoot there was a very little galena which, at least in part, occurred in quartz veinlets cutting the older quartz; it may belong to a distinctly later period, possibly related to the barite found in the flat fault.

In one locality only did the mineralization justify mining. This was from about 300 to 400 feet west of the shaft on the 300-foot level up to where the vein was cut off by the flat fault. Near the top, native silver was associated with the tetrahedrite. About

$65,000 worth of silver, copper, and gold was obtained from 1,200 to 1,300 tons of sorted shipping ore. Eighty per cent of the value was silver.

From near the Iron King tunnel to near the Shea shaft a quartz vein in the Copper Chief flat fault has largely replaced the gouge. This vein, which is several feet in maximum thickness, is similar to the Shea vein, although entirely barren quartz predominates to an even greater extent. Farther southwest only small stringers of quartz occur in the fault except at one spot, near the Grand Island shaft, where a short, thick lens produced over 100 tons of high-grade copper-gold ore. There was relatively little tetrahedrite, and the pyrite carried the gold.

Work in the flat fault vein more or less over the Shea ore shoot exposed a mass of at least several tons of barite in the flat vein and perhaps partly in the bent-over top of the Shea vein. The barite, which was not definitely associated with any other vein material, appeared to be younger.

ORIGIN OF PRIMARY MINERALIZATION

Conclusive evidence shows that practically all the Jerome district mineralization was replacement of pre-existing material by solutions. General knowledge of metalliferous ore deposits makes it practically certain that the source of the ore-bearing solutions was related to igneous rocks or igneous magma reservoirs. The particular igneous rock or inferred magma reservoir to which the ore solutions are related is speculative.

The location of the Jerome district with respect to the main mass of the Bradshaw granite and granitic material to the north and northwest makes the Bradshaw granite magma a very plausible deep-seated source for the solutions so far as location in space is concerned. It is reasonable to suppose that its magma may have been in an actively molten condition over a long period of time.

There is little reason to doubt that the vein mineralization of the district represents a late manifestation of the same mineralization that formed the massive sulphide bodies. The time location of the north-south granite-porphyry dikes between the vein formation and the earlier mineralization, together with the probable close affiliations of these dikes to the Bradshaw granite, indicates that the district mineralization was not widely separated in time from the intrusion of the granite.

Either the United Verde diorite or the Cleopatra quartz porphyry may very plausibly have been early differentiation products of the known Bradshaw granite. The diorite is similar to diorite closely associated with the granite in the quadrangle to the south.[27]

As the last important intrusive preceding the mineralization and because of its conspicuous association with the ore zone in the United Verde and U.V.X. mines, the United Verde diorite deserves

---

[27] T. A. Jaggar, Jr., and C. Palache, *U.S.G.S. Geol. Atlas*, Bradshaw Mountains Folio (No. 126), 1905.

first consideration. Otherwise there is little to suggest any genetic association. Away from the main ore zone the mineralization shows remarkably little tendency to favor diorite contacts. On the other hand, association of mineralization with the Cleopatra quartz porphyry is so widespread and conspicuous that it appears fairly certain that the association is more than a purely structural one, and in the main ore zone itself the evidence indicates that the first mineralizing solutions followed the porphyry contact more than the diorite contact.

The diorite does not occur very extensively in the district or surrounding area. The quartz porphyry is much more extensive, but probably its occurrence is largely within a radius of 5 miles from the center of the district.

It appears most probable that the ore-bearing solutions and the Cleopatra quartz porphyry had a common deep-seated origin in the Bradshaw granite magma.

GENERAL GEOLOGY

State of knowledge.—An article published by Provot[28] in 1916 correctly interprets the broader geological features of the district.

Reber's[29] 1920 paper was the result of more detailed study and observation. Dr. Lingren's[30] description, resulting from field work in 1922, is much more comprehensive and involves some corrections as well as important additions to the earlier papers.

Rickard's[31] early description of the U.V.X. Mine and Benedict and Fearing's[32] paper on the Verde Central Mine are of special local significance.

Even Dr. Lindgren's excellent and comprehensive description is subject to numerous minor corrections, as well as some very significant additions, in the light of all the information available to date. Hansen's[33] article outlines the most important of these additions.

It may never be possible to interpret some of the more obscure pre-Cambrian relationships with absolute certainty. Further detailed field work and much additional petrographic study are required. From a strictly practical viewpoint, however, the information now available is fairly adequate.

[28] F. A. Provot, "A Geological Reconnaissance of the Jerome District" (Arizona), Abs., *E.&M.J.*, CXXV (1916), 1028.

[29] L. E. Reber, Jr., "The Geology and Ore Deposits of the Jerome District" (Arizona), 1920, *T.A.I.M.E.*, LXVI (1922), 3-26.

[30] Waldemar Lindgren, *Ore Deposits of the Jerome and Bradshaw Mountains Quadrangles, Arizona* (U.S.G.S. Bull. 782, 1926), pp. 54-97.

[31] T. A. Rickard, "The Story of the U.V.X. Bonanza," *M. and S. P.*, CXVI (1918), No. 1, 9-17; No. 2, 47-52.

[32] J. L. Fearing, Jr., and P. C. Benedict, *Geology of the Verde Central Mine* (E. & M. J. Press, April 11, 1925), CXIX, 609-11.

[33] M. G. Hansen, "Geology and Ore Deposits of the United Verde Mine," *Mining Congress Journal*, XVI (April, 1930), No. 4, 306-10.

The following chronological outline, though necessarily in part somewhat speculative and subject to some valid differences of opinion, summarizes what is believed to be the best interpretation now possible.

CHRONOLOGICAL OUTLINE OF GEOLOGIC HISTORY

Pre-Cambrian.—1. Outpouring of a series of lava flows of widely varying composition, with some interbedded volcanic ash (tuff) and sedimentary material. Formation name, "Greenstone."

South end of district believed oldest because of prevalence of north to northwesterly dips throughout the district.

2. Outpouring of a great series predominantly of rhyolite lava flows and volcanic agglomerates of remarkably uniform composition, characterized by microscopic or very small visible quartz phenocrysts. Included with "Greenstone" on Plate VIII.

Because of the prevalently green color of the fresh rock and the difficulty often experienced in identifying the metamorphic material, the noncommital term, "greenstone," has been used for formations "1," "2," and "5," as well as "4" when not considered separately. It is also sometimes applied to material lacking the distinctive banding in "3." On the areal map, Plate VIII, "4" is shown separately.

3. Deposition of a series of volcanic tuffs and sedimentary material, for the most part distinctly bedded; at least some of the bedding is due to deposition in water. Formation name "Bedded Sediments." There were probably some contemporaneous lava flows.

In Haynes Gulch, north of the United Verde Mine, which is the type locality, there are some beds in which clastic sedimentary material undoubtedly predominates and one or two horizons showing well-rounded pebbles up to 1 inch in diameter. There is also at least one horizon containing from ½ to 2 feet of very typical "Lake Superior banded jaspilite or iron formation," which is much more extensively developed, in what is presumably the same horizon, south of the Yaeger Mine on the west side of the Black Hills.

3a. Period of regional deformation and folding with development of schistosity.

4. Intrusion of hornblende syenite, older diorite, and "Shea diabase."

The "older diorite," and hornblende syenite are not distinguished from the greenstone on the areal map (Pl. VIII).

5. Outpouring of a series of siliceous lava flows, first rhyolite, then predominantly dacite.

In addition to the large area shown on the map (Pl. VIII), small areas of this rhyolite occur as far north as the Copper Chief Mine.

5a. Some regional deformation.

6. Intrusion of Cleopatra quartz porphyry as a large continuous mass crossing the north end of the district, approximately on the greenstone-bedded sediments contact and a number of very ir-

regular intrusions forming a discontinuous belt crossing the south end of the district.

The area where porphyry predominates is shown as though it were a continuous belt (Pl. VIII).

The Cleopatra quartz porphyry is a rhyolite porphyry not very unlike the older rhyolites in composition. Its intrusive character and the distinctive appearance due to the prevalence of abundant large quartz phenocrysts as well as apparent significance in relation to the mineralization justify the distinctive name.

Ransome[34] has questioned the intrusive character of the north belt of porphyry. It is believed that evidence in the United Verde Mine indicating the intrusion of the bedded sediments by the porphyry is adequately conclusive in itself. It is fairly certain that all the greenstone along the south contact is surface volcanic material, and the nature of the contact is such that at least one of the rocks must be intrusive. The absence of any abrupt changes within the porphyry mass suggesting flow boundaries is also significant.

6a. Period of deformation, with somewhat local development of schistosity, and perhaps development of anticlinal structure more or less corresponding to area of general district mineralization.

The degree of schistosity in the porphyry varies greatly although most of it within the district is more or less schistose while such as known to the east of the mineralized area is relatively massive. This, as well as the prevalent N. 10 to 20 degrees W. trend of the schistosity in the northerly part of the district, is not inconsistent with the anticlinal structure suggested by the trend of several of the contacts and boundaries, but the prevalent east-west schistosity farther to the south is difficult to reconcile.

7. Intrusion of United Verde diorite.

8. General replacement deposition of fine-grained "jaspery quartz" by mineralizing solutions.

9. Somewhat widespread deposition of pyrite followed by less widespread deposition of brown sphalerite and probably more or less localized deposition of quartz carbonate.

10. Fairly widespread deposition of high-iron black chlorite.

11. Deposition of chalcopyrite with more or less pyrite followed by black sphalerite and galena locally.

12. Intrusion of numerous small andesite dikes, with prevalent east-west trend.

The composition of these dikes is very similar to that of the United Verde diorite.

13. Deposition of quartz followed by less widespread deposition of bornite.

14. Fairly widespread deposition of quartz, carbonate with a small amount of pyrite, and less widespread deposition of chalcopyrite, in part associated with light-colored sphalerite and very local tennantite.

---

[34] F. L. Ransome, unpublished report on United Verde Extension Mine.

15. Intrusion of Bradshaw granite.

This is hypothetical, although suggested by inferred relationship of north-south dikes.

16. Intrusion of granite-porphyry dikes along generally north-south dike zone.

These dikes are known only along a single rather definite belt or zone which has only been traced about 4 miles south of the Shea Mine but no doubt continues much farther. The actual trend is a little east of north to north of the Copper Chief Mine where it curves to a little north of east and then pinches out close to the Verde fault. Both microscopically and to the naked eye the rock from the larger dikes appears identical with a phase of the Bradshaw granite occurring to the southeast except for the absence of biotite and the local occurrence of hornblende phenocrysts.

17. Formation of quartz veins, typically with more or less coarsely intergrown ankerite, pyrite, chalcopyrite, tetrahedrite, and other minor constituents.

These veins generally follow premineral faults or other fractures, with a more or less east-west trend.

17a. Copper Chief thrust fault movement and perhaps also nearly horizontal movement on faults near Verde Central and very small thrusts in United Verde Mine.

18. Formation of additional quartz veins, more or less similar to 17.

19. Deposition of barite in the Shea Mine.

This, the only barite known in the district, may be younger than pre-Cambrian.

20. Extensive erosion with development of oxide and chalcocite zones in sulphide deposits.

21. Gentle submergence and deposition of youngest pre-Cambrian sedimentary formations.

This is purely by inference from the most probable age of mineralization.

21a. Gentle uplift followed by great faulting, with movement on the order of 2,400 feet vertical displacement along the Verde fault, which cut off and shifted the upper portion of the United Verde ore chimney to the northeast as well as downward.

22. Perhaps some pre-Cambrian erosion and further secondary enrichment of sulphide deposits.

Paleozoic.—1. Continued erosion and gradual submergence with perhaps some further secondary enrichment of sulphide deposits.

2. Deposition of great thickness of Paleozoic sediments interrupted by periods of uplift and erosion without perceptible tilting, initiated by the deposition of the basal Tapeats sandstone in Middle Cambrian time.

Mesozoic.—Either continuous erosion or some submergence and deposition of Mesozoic sediments which were later entirely removed.

Tertiary.—1. Continued erosion with gentle tilting so that Paleozoic sediments were completely removed from the Bradshaw Mountain area to the south, but a maximum thickness of nearly 1,000 feet was left close to Jerome and progressively greater thicknesses farther north.

1a. Development of early Tertiary Verde Valley with deep gulches near Jerome, one of which slightly penetrated the pre-Cambrian formations and partially exposed the top of the U.V.X. downfaulted segment of the Verde ore zone.

2. Interruption of drainage and filling of gulches with prebasalt gravels.

3. Outpouring of series of basalt flows, which totaled over 800 feet in thickness, with some interflow weathering and erosion.

4. Tertiary faulting. Development of the Tertiary Verde fault belt and gradual lowering of the Verde Valley area with contemporaneous filling of the valley with the Verde lake beds to a thickness of over 2,000 feet and contemporaneous gravel deposition from the fault scarps merging into the lake beds. Also some further outpouring of basalt during the lake-bed deposition.

Movement of the main Verde fault added to the separation of the two segments of the Verde ore zone and erosion of the fault scarp exposed the top of the United Verde segment to very active attack, resulting in the destruction of most of the pre-existing oxide and chalcocite zones. The redeposition of copper dissolved from the outcrop, in oxidized form in the limestone and in bouldery gravel in gulches below the fault, resulted in the formation of the Dundee deposit and the carbonate veins in U.V.X. ground near the Columbia shaft.

Quaternary.—Rejuvenation of drainage leading to present erosion cycle. Final draining of Verde Lake, dissection of lake beds, and more intense erosion of steep slope due to fault belt. Further erosion of top of United Verde ore zone and deposition of oxidized copper in basalt and prebasalt gravels in and along Verde fault and appreciable sooty chalcocite enrichment in U.V.X. ore zone. This secondary copper deposition was perhaps initiated during the preceding period.

Significant Dates in Human History

United Verde Mine and Jerome District.—

1875. Discovery of showings by U.S. Army scouts.
1876. Location of first mining claims.
1880-85. Working of United Verde gossan for silver.
1883. Operation of first copper smelter.
1888. Completion of railroad from Ashfork to Prescott. Supplies hauled 28 miles from Granite, near Prescott.
1889. Purchase of United Verde Mine by William A. Clark.
1894. First mine fire on 300 level of United Verde Mine. Completion of narrow-gauge railroad to Jerome.
1900. Start of fairly active prospecting of district.
1904-5. Operation of Equator Mining Company smelter in south end of district.

1914. Completion of standard-gauge railroad to Clarkdale.

1914-20. Very active prospecting throughout district.

1915. Removal of United Verde smelting operations from above the mine to new plant at Clarkdale.

1916. Copper Chief 125-ton cyanide mill in operation. Finally closed down in 1923.

1919. Beginning of United Verde open-pit operation.

1920. Completion of standard-gauge railroad to Jerome.

1921. Completion of direct highway from Jerome to Prescott.

1924-27. Series of large "coyote" blasts in open-pit stripping.

1927. Addition of flotation concentrator to Clarkdale plant.

1929. January—Verde Central 350-ton flotation mill in operation. Closed in 1930. July—Cave-in of upper levels of United Verde Mine. United Verde peak copper production—142,290,000 pounds of copper from 1,737,000 tons of ore, for year.

1931. Purchase of Verde Central property by United Verde.

1935. Purchase of United Verde property by Phelps Dodge Corporation.

1938. United Verde open-pit work nearing completion.

United Verde Extension.—

About 1900. Verde Queen smelter treating carbonate ore from near Columbia shaft. (Later U.V.X. property.)

1900. Location of Little Daisy fraction by J. J. Fisher, the first U.V.X. claim.

1902. Organization of U.V.X. Company.

1912. Reorganization of U.V.X. Company by J. S. Douglas and G. E. Tener.

1914. First important ore discovery in U.V.X. Mine.

1915. Discovery of main ore body.

1917. U.V.X. peak copper production—63,243,000 pounds of copper from 115,064 tons, for year.
First production from Jerome Verde, main top ore body (mined out in 1920).

1918. Completion of U.V.X. smelter at Clemenceau.

1929. U.V.X. peak tonnage production—59,127,000 pounds of copper from 358,650 tons of ore, for year.

1930. Addition of 200-ton flotation mill to plant at Clemenceau.

1937. January—U.V.X. smelter permanently closed.

1938. May—U.V.X. Mine permanently closed.

Acknowledgments

The writer wishes to acknowledge the courtesy of the officials of the Phelps Dodge Corporation in permitting the publication of this paper and the use of United Verde material, and of J. S. Douglas in permitting the use of U.V.X material.

MIAMI-INSPIRATION DISTRICT[35]

By G. R. RUBLY[36]

HISTORY AND PRODUCTION

The major developments in the Miami-Inspiration district have been of comparatively recent date. At the beginning of the century, chrysocolla had been mined at the Keystone Mine, and soon after a vein of chrysocolla was stoped at the Live Oak. Both of the veins were in granite porphyry and did not extend into the schist. Several years later the Woodson tunnel was driven in the north side of Inspiration Ridge. This tunnel cut disseminated chalcocite, and some crude ore was mined from a zone of stringers in the schist. In 1906 the General Development Company sank a shaft on the Captain claim and another on the Red Rock, the latter striking ore at a depth of 220 feet. The Miami Copper Company was organized in November of that year and development work was actively undertaken. By 1909 the railroad had been extended to Miami from Globe, and in 1911 the first concentrates were produced after an intensive construction period which saw the completion of a mill, power plant, and other surface equipment.

During this period the Inspiration Copper Company and the Live Oak Development Company were also engaged in development work. At Inspiration active development by shafts, drifts, and crosscuts, as well as churn drilling, was begun in 1909. Two years later, 21,000,000 tons of ore had been outlined.

The Live Oak Development Company had by 1912 developed 15,000,000 tons of ore despite the fact that much of the ore body lies deeper than at either Miami or Inspiration and is covered by porphyry and Gila conglomerate. The Live Oak Development Company and Inspiration Copper Company consolidated in January, 1912, as the Inspiration Consolidated Copper Company with ore reserves of 45,300,000 tons averaging 2 per cent copper.

Further development and refinements in mining and milling methods have since greatly increased the ore reserves of both major companies.

The following production[37] has been recorded:

	Pounds of copper	Total value
Inspiration, 1905-32	1,336,500,000	$227,500,000
Miami, 1911-33	1,072,500,000	171,500,000
		$339,000,000

---

[35] Paper obtained for, and originally presented at, the regional meeting of the A.I.M.&M.E. held at Tucson, Arizona, November 1-5, 1938.

[36] Chief Mine Engineer, Miami Copper Company.

[37] M. J. Elsing, and R. E. S. Heineman, *Arizona Metal Production* (Univ. of Ariz., Ariz. Bureau of Mines Bull. 140, 1936), p. 92.

Rocks

Although several geological formations occur in the district, comparatively few are immediately connected with the ore deposits of Miami and Inspiration. They may be enumerated as the Pinal schist, Pioneer shale, diabase, Willow Spring granite, Schultze granite, dacite, and Gila conglomerate.

The oldest formation, one of the host rocks of the ore deposits, is the Pinal schist of pre-Cambrian, possibly Archean age. The mass of the Pinal Range, which covers an area about 16 miles long from northeast to southwest by about 12 miles wide, is made up largely of Pinal schist with considerable irregularly intruded pre-Cambrian quartz diorite and granite and also a younger intrusive, the Schultze granite. The disseminated copper deposits of Miami occur in the northeast corner of this area.

The schist folia are generally contorted, and the dip of the schistose cleavage ranges from 45 degrees to vertical. The prevailing strike is northeast, in general not parallel to the present mountain ranges but nearly at right angles to them.

The Pinal schist is made up principally of a series of metamorphosed quartzites and shales. The effects of the intrusion of pre-Cambrian quartz diorite and granites are so pronounced near the contacts and decrease so gradually outward as to lead to the conclusion that the first crystalline metamorphism dates from these intrusions.

One of the widespread and important granitoid rocks, known as the Madera diorite, is most abundant in the Pinal Range where it is intricately intruded into the Pinal schist. Its effects on the schist are important to the mining of copper ores in the district.

Of all the Paleozoic sedimentary rocks of the region, only the Scanlan conglomerate and basal portions of the Pioneer shale remain.

Because of the great disturbances that accompany its intrusion, the diabase is believed to have reached the surface. No effusive rocks of this period now remain in the region. The magma forced its way into the sedimentary strata as sheets and sills, but these are not persistent, as it also broke across the bedding planes in so many places and followed so many different planes of stratification that the resulting structure is highly irregular, and in places great masses of sediments are entirely surrounded by diabase. This same rock in places penetrated the Pinal schist as dikes and sills, many of which carry chrysocolla and native copper.

The Willow Spring granite forms a small mass bounded by faults just north of Webster Gulch.

Probably the most important rock associated with the copper deposits of Inspiration and Miami is the Schultze granite. It occurs as an irregular body extending from the vicinity of Miami on the northeast to the Pinal Ranch on the southwest, a distance of about 10 miles. It is roughly divided into two parts by a constriction northwest of Bloody Tanks. The northern lobe is largely porphyritic. This granitic porphyry has been minutely fissured and the cracks have been filled with quartz and to a less extent with sulphides. The result is a very brittle and fragile mass that, with the crumbly character of the Pinal schist, is so essential to successful block-caving methods of mining.

Dacite originally covered most of the Globe and Ray quadrangles, but its continuity has been greatly decreased by erosion and faulting. Although its maximum thickness is unknown, remnants indicate a thickness in excess of 1,000 feet.

While data are lacking for fixing the date of eruption of the dacite, it is known to have occurred after the eruption of granitic, monzonitic, and dioritic rocks and undoubtedly preceded the deposition of the Gila conglomerate. It is probably of Tertiary age.

The accumulation of the Gila conglomerate is indicative of intensely active erosion following the period of vigorous deformation that outlined the present mountains and valleys of the region. Postdacitic earth movements apparently threw the land into the high relief which with the semiarid climate was so favorable to rapid mechanical disintegration and erosion. The thickness of the deposit varies greatly from place to place and perhaps exceeds 8,000 feet.

No workable deposits occur in the Gila conglomerate, but it covers a portion of the ore bodies and must in places be penetrated to reach the metallized schist below.

STRUCTURE

The Miami district contains numerous faults which are probably not all of the same age, and which are definitely not amenable to systematic grouping. These numerous faults, both prior to and later than metallization, contribute much to the ease of working the ore bodies.

The most conspicuous break is the Miami fault which limits the ore of the Miami Mine on the east. This fault in general strikes N. 20 to 30 degrees E. and dips at an average of 45 to 50 degrees SE. It is a normal fault with Gila conglomerate faulted mainly against Pinal schist. Its throw has not been determined.

On Inspiration ground the Joe Bush fault can be indistinctly traced along the surface, and its underground exposures are conspicuous. For a distance of about 1,500 feet it separates barren porphyry on the southwest from ore-bearing schist on the northeast.

The Bulldog fault is a zone of strong faulting with a course nearly north and a low angle eastward. The effect of this zone of faulting has been to displace the ore body so that the ore of the Inspiration Mine on the east is 450 to 500 feet lower than the ore of the Keystone on the west.

The faulting in the vicinity of the Live Oak No. 2 shaft is structurally important but has some very puzzling features. It is not within the scope of this paper to do more than mention this faulting.

Another displacement of interest is the Pinto fault which drops the dacite against the schist on Inspiration ground and the schist against the porphyry on Miami ground.

Ore Deposits

The metallic minerals of interest occurring in the Miami-Inspiration district are native copper, native silver, molybdenite, galena, chalcocite, covellite, chalcopyrite, pyrite, cuprite, malachite, azurite, and chrysocolla.

Three terms—protore, ore, and capping—are used in referring to the metallized rocks that make up the ore bodies of the district.

Protore is used to designate the metallized rock of a grade or tenor too low to be classed as ore but which would have been converted to ore had the enriching cycle been carried sufficiently far.

Ore may be considered as material that can be mined and treated at a profit under favorable price conditions.

Capping refers to the leached, practically barren overburden.

Examination of almost any specimen of ore shows that the separate particles of copper minerals are not disseminated evenly through the mass of rock but to a large extent lie in minute cracks and joints. The ore tends to break along these joints so that fracture faces show more ore mineral than a saw cut across the rock. The term "disseminated," while being used almost universally in referring to ore bodies of this type, is not strictly correct, but its use is not objectionable if the fact is kept in mind that the mineral particles are not scattered entirely through the rock.

The bodies of disseminated copper ore of Miami and Inspiration may be characterized generally as undulating, flat-lying masses of irregular horizontal outline and variable thickness (Fig. 6). As a rule these masses lack definite boundaries. There are no easily recognizable distinctions in color, texture, or general appearance to mark them off from the enclosing rocks, and closely spaced sampling and assays indicate a gradational passage from ore to country rock. Likewise the size and shape of the ore bodies depend largely on the local and current definition of ore. As a rule the transition downward from leached and oxidized capping to ore is less gradual than that between ore and protore.

The depth to the ore ranges greatly from place to place, as in many places the leached rock itself is overlain by dacite or Gila conglomerate. Drill holes show maximum depths of 1,100 feet which in no way are related to the present topography.

In the Miami district the lowest ore occurs where the surface rocks consist of dacite or Gila conglomerate which, it might be supposed, would have tended to prevent oxidation and enrichment of the underlying sulphides. This relation inevitably leads to the conclusion that enrichment took place prior to eruption of the dacite and deposition of the conglomerate.

The ore shows no regular relation to the ground-water surface except that the ore generally lies deeper to the west, and the water surface goes deeper to the east. This irregularity indicates

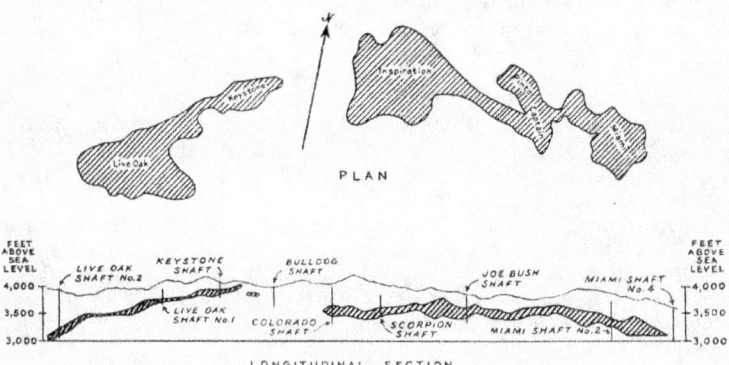

Figure 6.—Generalized plan and longitudinal section of developed, disseminated ore bodies in Miami district.   (After Ransome, U.S.G.S. Prof. Paper 115.)

that the enrichment processes which produced the ore bodies took place before the development of the present topography.

In general the bottom of the oxidized zone is a fairly definite but uneven and undulating surface. Oxidation appears to have been limited by the gradual exhaustion of oxygen from air and rain water as they penetrated the formations, rather than a coincidence with a ground-water surface. This being the case, oxidation would extend deeper along certain favorable channels through the mass of rock, and as such permeable zones would usually not be vertical it is easy to account for the occurrence of oxidized material beneath enriched sulphide ore and protore. Such an occurrence is encountered in the Miami Mine.

As already noted, the rocks intimately associated with the disseminated ores in the Miami district are the Pinal schist and the Schultze granite. The greater part of the ore is metallized schist, whereas a relatively small amount is metallized granite porphyry.

The igneous rock cannot be regarded only as material that has in part been converted to ore but must also be considered as a possible active agent, through the consequences of intrusion, in the process of metallization.

In a general way the ore in the Miami district occupies a marginal position with reference to the porphyritic facies of the Schultze granite. It does not, however, follow the contact closely nor is it entirely in one rock. The Miami ore body is chiefly in schist, although a granite-porphyry dike cutting the schist has likewise been metallized. The Inspiration ore body is also mainly in the schist, although in places this schist occurs beneath an intrusive sheet of the porphyry.

Schist and porphyry have both been susceptible to metallization, and there is no essential difference between ore in schist and ore in porphyry even at the contact, except where such contact is due to a fault. There may be an abrupt change in character of the ore at such a fault contact, just as there might be in passing a fissure wholly in schist or wholly in porphyry.

It is fair to conclude, then, that the visible portions of the porphyry mass near the ore bodies were not sources of metallization but, like the schist, were acted upon by metallizing solutions originating in some more distant and probably deep-seated mass of rock.

No evidence has been found to indicate that any of the faults exposed in the Miami district are older than the hypogene metallization. That some of them may have been existant prior to enrichment is difficult to establish, for while it is generally believed that most of the supergene enrichment took place before the development of the present topography, the process has undoubtedly been carried on up to the present day. As a result a fault may antedate some enrichment and yet postdate a great deal of it. The occurrence in the gouge of rounded fragments of chalcocite ore and crushed oxidized material evidently derived from the leached rock overlying the ore is fair evidence that faulting generally postdated the beginning of enrichment. The Miami fault and some faults near the Live Oak Mine cut dacite and conglomerate, formations which are younger than most of the enrichment.

The close association of granitic and monzonitic porphyries to ore bodies is well known. It may be safely concluded that they had something to do with the formation of these ore bodies. It is not to be taken for granted, however, that the now visible parts of these porphyries contributed in any active way to metallization, for, they like the schist, have also been altered by the ore-bearing solutions, and, where situated favorably, have been changed into protore just as was the schist when so situated. They merely testify to the probable presence of much larger masses of similar igneous material far below any depths likely to be reached by mining, which must have taken much longer to solidify and cool than the now exposed portions and from which most of the energy and part of the materials of metallization emanated.

It is generally accepted that the deposits of the district are due to the action of thermal alkali sulphide waters probably carrying carbon dioxide. Two elements that the ascending solutions brought up from below were copper and sulphur with probably molybdenum and silicon. Comparative analyses on schist samples of equal weight show some silicification. The iron already present in the silicates of the schist and porphyry and as magnetite was more than sufficient to form the pyrite now in the rock.

The hypogene solutions, on the whole, appear to have been of weak chemical activity, and, although the quantity of copper transported was enormous, it was small in terms of percentage of the rock mass.

There are no clear relations between primary metallization and structure. Typical protore is found more than 1,000 feet from any known porphyry mass of considerable size. Permeability of the rock mass and an abundant supply of active solutions were probably more essential to metallization than any other combination of factors. Permiability was due largely to the minute fissuring.

Deposition of protore probably followed closely the intrusion of the granite porphyry, but no known facts are available to fix this event in geologic time. It is reasonable to assume that the granite porphyry was intruded in Laramide time, certainly after deposition of the Mississippian and Pennsylvanian limestones and before eruption of the dacite.

The early stages of the enrichment must have been very slow. Under normal conditions the pyritized rock would be brought within reach of supergene solutions only after erosional removal of the overlying rock. The sulphides first reached probably were sparsely disseminated, and oxidation proceeded only here and there where conditions were especially favorable. As successively deeper portions were attacked, chemical activity would increase, and downward movement and concentration of copper would be in full progress.

Oxidizing solutions first attacked pyrite and chalcopyrite. Increasing quantities of chalcocite were deposited lower down at the expense of pyrite and chalcopyrite, and, as erosion progressed, this chalcocite came into the zone of oxidation. A stage must have been reached in which chalcocite and not pyrite was the object of attack of the descending atmospheric waters. When this stage is reached the process slows down so that if the erosional rate remains the same or is accelerated, it may overtake enrichment and attack rock that contains copper carbonate and silicate formed above the chalcocite.

The process of enrichment is potentially a cyclic operation, but many complicating factors, such as rate of erosion, depth to underground water, climate, topography, and country rock influence it.

REFERENCE

Ransome, F. L., The Copper Deposits of Ray and Miami, Arizona, U.S. Geol. Survey Prof. Paper 115, 1919.

CLIFTON-MORENCI DISTRICT[38]

By

B. S. BUTLER AND ELDRED D. WILSON

INTRODUCTION

The following general description of the Clifton-Morenci district is summarized from Dr. Lindgren's classic report,[39] published in

[38] Paper obtained for, and originally presented at, the regional meeting of the A.I.M.&M.E. held at Tucson, Arizona, November 1-5, 1938.
[39] Waldemar Lindgren, *The Copper Deposits of the Clifton-Morenci District, Arizona* (U.S. Geol. Survey Professional Paper 43, 1905).

1905, and the more detailed description of the Morenci open-pit mine is mainly from the records of the Phelps Dodge Corporation. The writers are particularly indebted to W. C. Lawson, Chief Engineer, for information.

In 1901 and 1902, when Dr. Lindgren studied the district, the mining of the low-grade ore in the monzonite porphyry of Copper Mountain was already well advanced. The low-grade ore then mined was much richer than the disseminated ore later so extensively mined in the Southwest but was of the same general type. Its successful exploitation at Morenci pointed the way to the development of other great disseminated deposits.

Dr. Lindgren's report was the first detailed description of a disseminated copper deposit, though of a somewhat specialized type, and it, too, pointed the way to the development of the disseminated deposits.

In the third of a century since the publication of Professional Paper 43, advancement in the mining and metallurgy of copper ores has steadily lowered the grade that can be profitably mined until material then of no value has risen to large value. Such is true of part of the mineralized area of Morenci that is now being developed into one of the largest copper mines of the state. Thus, the Morenci district, which was first in the utilization of disseminated copper ores, is now developing the latest mine in this type of deposit.

The main purpose of the present paper is to show the relation of this disseminated deposit to the other deposits of the district and to give a brief description of the geology of the district.

HISTORY

Some of the important events of the district with dates follow:

1872. Discovery by the Metcalf brothers.
1873. Sale of important claims to the Lesinskys.
    Organization of Longfellow Copper Company by Lesinskys and building of adobe furnace.
1875. Detroit Copper Company organized.
1882. Sale of Longfellow Company to Arizona Copper Company.
1887. Phelps, Dodge, & Company acquire interest in Detroit Copper Company.
1893. Discovery of low-grade sulphide ores of Copper Mountain.
1899. Organization of Shannon Copper Company.
1919. Purchase of Shannon Copper Company by Arizona Copper Company.
1921. Purchase of Arizona Copper Company by Phelps, Dodge, & Company.
1937. Beginning of Morenci open-pit operations.

ROCKS

The oldest rock of the district is pre-Cambrian granite. Resting on the granite is the Cambrian (Coronado) quartzite, 200 feet thick, followed by the Ordovician (Longfellow) limestone, 380 feet; the Devonian (Morenci) shale, 175 feet; and the Carboniferous (Modoc) limestone, 170 feet.

Unconformably on the Paleozoic rocks are a few hundred feet of Cretaceous shale and sandstone.

The Cretaceous and earlier rocks are intruded by a stocklike body, porphyritic in texture and ranging in composition from granite through monzonite to diorite. Extending from the stock into the surrounding rocks are dikes, sills, and laccoliths. Following a period of erosion that removed the cover from the intrusive bodies and exposed the copper deposits to oxidation, the area was buried beneath volcanic flows, breccias, and tuffs, ranging in composition from basalt to rhyolite.

Following the outpouring of volcanic material, the lower areas were buried by coarse, poorly assorted gravels (the Gila conglomerate) deposited by streams from the higher ground. In the Clifton area the Gila conglomerate has been rather deeply trenched and redeposited as bench and stream gravels.

STRUCTURE

The region doubtless had a complex structural history in pre-Cambrian time, but the pre-Cambrian rock is largely granite and reveals little of its early structural history.

From Cambrian time to the close of Cretaceous, the area alternated between erosion and deposition, the result of broad uplift and depression, but with no pronounced folding or faulting recognized.

At or near the close of Cretaceous, stresses developed that fractured the rocks in a generally northeast-southwest direction. Into this fracture zone magma rose to form the porphyry stock and associated dikes, sills, and laccoliths.

Along the contact of the porphyry stock with the sedimentary rocks, large and small masses of the earlier rocks were engulfed in the porphyry (Pl. XVI).

Stresses of the type that produced the breaks along which the magma rose continued after its solidification and produced a series of prominent northeasterly striking fissures in the porphyry. Such stresses doubtless caused continued movement on the earlier breaks and opened new breaks in the older rocks. These are the mineralized fissures that strike from north to east, with a prevailing direction of about N. 35 degrees E.

Associated with the northeasterly fissures are less persistent breaks that seem without rhyme or reason. If mapped in detail, they would doubtless fall into definite systems. These fractures broke the rocks, especially the porphyry, into small fragments bounded on all sides by fissures (Pl. XVII).

Some of the more prominent northeast fissures, including the Humboldt and the Wellington in Copper Mountain, and fissures in the open-pit area are shown on Plate XVI.

Especially prominent near Morenci are northwesterly striking

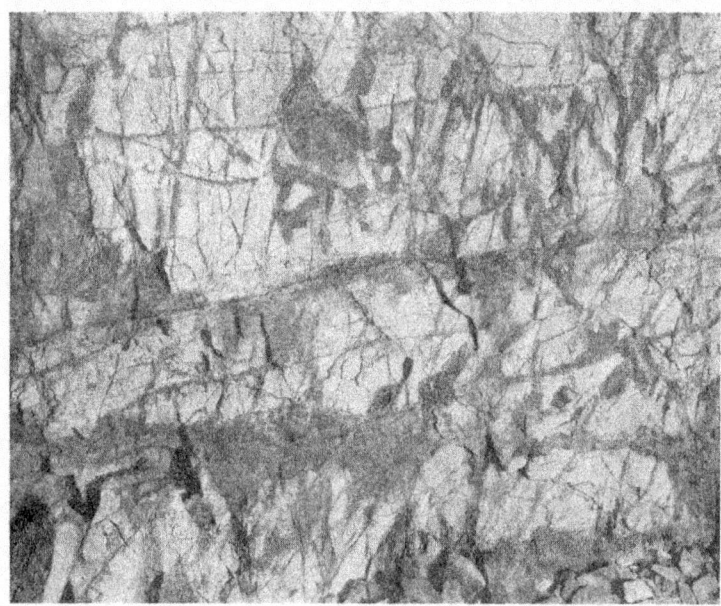

Plate XVII.—View of shattered monzonite porphyry, fifth underground level, open-pit area. (Photograph, courtesy Phelps Dodge Corporation.)

faults that determine the location of some of the principal gulches. These faults are later than the mineralization of the northeast fissures and associated breaks and even later than the enrichment of the veins by oxidation processes but earlier than outpouring of the lavas. It seems probable, however, that these breaks were initiated before the intrusion of the monzonite porphyry.

ORE DEPOSITS

The primary mineralization was probably all at essentially one time, closely following the intrusion and solidification of the porphyry and the fissuring of it and the older rocks.

The deposits may be separated into three main groups: (1) those that replaced sedimentary rocks, (2) those that formed in fissures and as disseminations in the monzonite porphyry, and (3) fault-fissure veins of the Coronado type.

DEPOSITS REPLACING SEDIMENTARY ROCKS

The early operations were largely on oxidized ores of deposits that replaced sedimentary rocks. The primary ores consisted of pyrite, chalcopyrite, and sphalerite in a gangue that contained abundant silicates of which garnet and epidote are most abundant. These minerals replaced favorable strata of the limestone with material of low copper content. The ore resulted from the oxidation of this material with the change of the copper minerals from sulphides to carbonates and oxides and with an enrichment in copper.

The Longfellow Mine at Morenci is of this type as are numerous other deposits at both Morenci and Metcalf. Production from this type of deposit has been worth several million dollars.

Fissures and Disseminations in Monzonite Porphyry

Due to the barren outcrops of the veins in porphyry which had been leached to a depth of some 200 feet, the deposits in the monzonite porphyry remained undiscovered for several years after the development of the deposits in the limestones. The early operations were confined to the larger veins, but as it became possible to treat lower grade material, more and more of the wall rock was mined. Lindgren[40] wrote in 1905:

On both sides of these seams extends a mass of sericitized porphyry, of varying width, containing little seams and grains of chalcocite. To this ore there are no well-defined walls, but they gradually fade out into material too lean to constitute ore. Could two percent ore be made payable, the width of the ore bodies would be much increased.

When it became profitable not only to treat 2-per-cent ore but material of much lower grade, the width of the ore bodies was so extended that it became practicable to take all the material between fissures and to mine the lode zones as one ore body.

Professor Lindgren[41] describes the vein of Copper Mountain as follows:

The most important vein system is that which, under the general name of the Humboldt vein, extends from northeast to southwest through Copper Mountain at Morenci. The outcrops of this vein are practically barren, but at a depth of 200 feet the deposit becomes productive and contains chalcocite associated with pyrite. There are usually one or more central seams of massive chalcocite, some of which are fairly persistent. These seams are ordinarily adjoined by decomposed porphyry, now chiefly consisting of sericite and quartz, together with pyrite and chalcocite. These extensive impregnations of the country rock are rarely confined to distinct walls, but gradually fade into the surrounding porphyry. That these deposits are genetically connected with fissure veins, however, cannot be doubted. In lower levels the ore is apt to change to pyrite and chalcopyrite.

This, as a general description, applies equally well to the more recently developed deposits, including the open-pit mine.

Characteristic features.—The lode systems have a general northeast trend. Mineralization is strong in the main fissures and ex-

[40] *Op. cit.*
[41] *Op. cit.*

tends into the smaller associated fissures. The primary mineralization, as shown in the deep levels, is quartz, pyrite, chalcopyrite, and sphalerite veins with most veins rich in pyrite. Outcrops over the ore are largely leached of both iron and copper. Ore croppings are light buff in color as contrasted with the deep red and brownish red tones common in mineralized porphyry that does not contain ore.

Below the leached cap rock, which has a maximum thickness of 500 feet, the ore consists of secondary chalcocite that has replaced pyrite extensively in the rich ore and decreasingly with increased depth or decreased permeability of the rock.

Development usually stops where the copper content falls below the commercial grade, though the copper may still be present in part as secondary chalcocite. There are, therefore, no very definite data on the copper content of the primary mineralization, but it is low and the ore represents a relatively large addition of copper from the leached zone.

The present water table is below the ore that has been enriched, and the irregular lower surface of the ore indicates that it has not been controlled by a water table. Much of the enrichment was accomplished before the burial of the area by lavas, and the water table may then have had a very different relation to enriched ore. Since erosion of the lava from above the deposits, the enrichment process has been renewed.

Coronado Vein

The Coronado vein occurs as a cementation of breccia on the Coronado fault which strikes east-northeastward and has thrown Coronado quartzite against pre-Cambrian granite. A diabase dike was intruded into this fault before movement had ceased and before mineralization. The primary mineralization was similar to that of the fissures in the monzonite porphyry, but alteration resulted in the formation of a zone of oxidized ores and one of secondary sulphide ores. This contrasts with the deposits in limestone which were mostly oxidized ores and those in the monzonite porphyry which were mostly sulphide ores. The difference is probably due to the different types of rock in the different deposits. The reactive limestone yielded oxidized deposits, the relatively inert monzonite porphyry yielded secondary chalcocite, and the intermediate rocks of the Coronado vein yielded both.

The Morenci Open-Pit Mine

Location.—The Morenci open-pit mine is on the southwest side of Chase Creek and north of the area that has been most productive in the past—namely, the Copper Mountain-Longfellow area (Pl. XVI).

The ore body, as now defined, has a maximum length of 4,400 feet and a maximum width of 2,800 feet.

Rocks.—The mine is entirely within the quartz monzonite porphyry body, though there are some blocks of mineralized quartzite engulfed in the porphyry within the mine. The monzonite is highly altered throughout the mineralized area. No detailed study of the alteration of this area has been attempted, but it is clearly the same in all essentials as that of the Copper Mountain area described by Lindgren,[42] and by Reber.[43] Typically, it consists of sericitization together with the introduction of pyrite and chalcopyrite.

Structure.—The most striking feature of structure is the fissuring. In most faces, fissures seemingly without definite order or pattern break the rocks into small angular fragments (Pl. XVII). A mapping of the more prominent fissures, however, indicates very definite directions.

Different parts of the ore body have been developed by different methods. Some of the earliest openings followed the more prominent fissures for hundreds of feet. Later, areas were developed by underground openings laid out in a rectangular pattern, and the latest developments were by core drill. The conspicuous fractures in the underground openings are shown on the Company maps. No attempt at a statistical compilation of direction has been made, and it is difficult to evaluate the strength of the fissures from the records. A general inspection of the maps, however, gives the impression that fully three fourths of the mapped fissures strike within the northeast quadrant and that those within this quadrant are predominantly about northeast.

In the part of the mine early developed along the most prominent fissures, the northeast direction is particularly conspicuous. Some strike nearly north. The fissures for a part of the fifth level are shown on the plan map of the ore body (Pl. XVIII). Similar fissures are doubtless present in other parts of the ore body, but the data for plotting them are not now available. The fissures that were sufficiently prominent to induce vein prospecting are widely spread, 300 to 400 feet apart, but between them are networks of smaller fissures that, as already stated, break the whole body of the rock into small fragments.

Ore body.—It may be noted that the greatest dimension of the ore body is in the direction of the strike of the more prominent fissures—namely, northeast.

Lindgren has shown that the disseminated deposits of the Copper Mountain area are unquestionably associated with the northeast fissures of that area. It seems equally evident that the disseminated deposits of the open-pit mine are associated with northeast fissures.

In the estimation of ore reserves, assay maps were prepared of all openings. As these maps clearly show, drifts that follow northeast fissures have much higher average copper content than openings laid out on a rectangular pattern, or those that do not follow fissures.

---

[42] Waldemar Lindgren, *Op. cit.*

[43] L. E. Reber, Jr., "Mineralization at Clifton-Morenci," *Econ. Geol.*, XI (1916), 528-73.

No attempt has been made to ascertain the copper content of fissure drifts, but they evidently average two to three times that of the average of the ore body. If the fissures alone were sampled instead of the whole width of the fissure drift, the average would be much higher.

The concentration in the larger fissures is so evident that in estimating copper content of the ore body, the fissure drifts were not included.

The mineralized body, in common with other disseminated deposits, can be separated into three parts or zones: (1) the upper or surface zone, (2) the ore zone, and (3) the lean sulphide zone.

1. The upper or surface zone consists of iron-stained, silicified, and sericitized monzonite. Over the ore body this is generally of light buff color with darker red to brown streaks along the fissures in contrast with much higher colored red and brown capping of the rocks surrounding the ore body.

The copper content of the capping rock as a whole is not shown on the assay plans, but the few recorded assays and the visible features of the rock indicate that, near the surface, the removal of the copper has been nearly complete.

The oxidized capping has a maximum thickness of 500 feet, with an average of 216 feet. In general, it is thickest under hills and thinnest under valleys.

In places, the lower part of the oxidized zone contains considerable copper so that there is not everywhere a sharp change in copper content between the oxidized zone and the sulphide zone.

2. The ore zone lies beneath the oxidized zone. Its blue-gray color contrasts sharply with the buff capping rock. At closer range, the network of fissures becomes prominent (Pl. XVII). Examination shows the fissures to be composed mainly of quartz and pyrite, the latter coated and replaced to varying degrees by chalcocite.

The thickness of the ore body is irregular, but over much of the area is 500 to 700 feet (Pls. XIX and XX). There is a general pitch of the ore zone eastward in the direction of greatest elongation. The bottom of the ore to the east is some 200 feet lower than to the west. The same in general holds for the top of the ore, and both roughly correspond with the slope of the present erosion surface.

There are no sharp boundaries to the ore either laterally or in depth. The boundaries as shown are based on expectation of profitable extraction. In depth, however, many drill holes show a rather sharp drop from near 1 per cent to ½ per cent or less. As most drill holes were stopped when the copper fell to ½ per cent, there is no general record of the copper content of the underlying material. Some holes, carried 200 feet or more below the ore, indicate that the unenriched material contains less, perhaps considerably less, than ½ per cent copper. In this material the copper is probably in chalcopyrite.

The zone of secondary enrichment, when formed, before the outpouring of the later lavas, may have had some definite relation to a water table, but neither the top nor the bottom of the secondary enriched zone is now determined by a water table.

The Phelps Dodge Corporation prospectus covering the issue of convertible 3½ per cent debenture bonds, in regard to the Morenci open-pit ore reserve, estimates 284,000,000 tons of ore assaying 1.036 per cent copper. The ore carries small and relatively unimportant amounts of gold and silver. The ore available for extraction on the basis of the pit lay-out now contemplated for this program is estimated at 230,000,000 tons carrying 1.06 per cent copper.

BIBLIOGRAPHY

Lindgren, Waldemar, The Copper Deposits of the Clifton-Morenci District, Arizona, U.S. Geol. Survey Prof. Paper 43, 1905.
  U.S. Geol. Survey Geol. Atlas, Clifton Folio (No. 129), 1905.
Reber, L. E., Jr., The Mineralization at Clifton-Morenci, Arizona, Econ. Geology, Vol. 11, pp. 528-73, 1916.
Tenney, J. B., Copper Deposits of Morenci District, XVI Int. Geol. Congress, Copper Resources of the World, Vol. I, pp. 213-21, 1935.

RAY DISTRICT[44]

INTRODUCTION

Ray is about 17 miles south of Miami on Mineral Creek, between the Dripping Spring Range on the east and the Tortilla Mountains on the west.

The latest and most complete report on the Ray district is by Dr. F. L. Ransome.[45] Since the time of Dr. Ransome's report, underground development has further revealed the extent of faulting and the structural relationships.

The present paper gives a general description of the geology largely summarized from Ransome's reports, together with new data regarding the structural features of the ore body. These new data, together with Figure 7, are taken from an unpublished manuscript by C. Leroy Hoyt.[46]

HISTORY AND PRODUCTION

1873. Mineral Creek district organized prior to this time by silver prospectors.

---

[44] Paper compiled for the regional meeting of the A.I.M.&M.E. held at Tucson, Arizona, November 1-5, 1938.

[45] F. L. Ransome, *Copper Deposits of Ray and Miami, Arizona* (U.S. Geol. Survey Prof. Paper 115, 1919); *Description of the Ray Quadrangle* (U.S. Geol. Survey Folio 217, 1923).

[46] Engineer, Nevada Consolidated Copper Company.

1880. Location and prospecting of claims. Mineral Creek Mining Company builds a five-stamp mill.
1883. Ray Copper Company organized with a capital of $5,000,000.
1898. Small-scale operations at Ray Mine.
1899. Ray Mine acquired by Ray Copper Mines, Ltd., capitalized at £260,000. Building of 250-ton mill at Kelvin and blocking out of ore at mine. This company failed because of inadequate sampling.
1907. Ray Consolidated Copper Company organized by D. C. Jackling and others to acquire and work the ground formerly held by the Ray Copper Mines, Ltd. Arizona Hercules Copper Mining Company and Kelvin-Calumet Mining Company begin operations.
1909. The existence of about 50,000,000 tons of ore is ascertained. Ray Central Copper Mining Company succeeds Kelvin-Calumet Mining Company.
1910. Louis S. Cates becomes Superintendent of Mines and develops mining system whereby Ray later became the first copper mine in the world to produce 8,000 tons or more of ore per day by caving methods.[47]
1911. Production starts from mines of Ray Consolidated Copper Company after construction of mill at Hayden.
1924. Ray Consolidated and Chino companies merge.
1926. Nevada Consolidated absorbs Ray and Chino.

Production of the Ray mines to the end of 1931 has been recorded[48] as follows:

TABLE 4.

	Copper (pounds)	Gold (value)	Silver (value)	Total value
Ray, 1911-31	1,156,000,000	$280,000	$150,000	$198,500,000
Ray Hercules, 1918-23	8,000,000			1,500,000
Total	1,164,000,000	$280,000	$150,000	$200,000,000

ROCKS[49]

The oldest rocks in the region are the Pinal schist, which consists mainly of metamorphosed siliceous sediments and various granitic intrusive rocks. All these rocks are of older pre-Cambrian age. Resting on the eroded surface of the old crystalline rocks are Apache group beds (Pl. III) aggregating from 1,200 to 1,300 feet in thickness, apparently in conformable sequence and supposed to be younger pre-Cambrian. More than two thirds of this thickness is represented by two quartzite formations; the remaining beds include shale, dolomitic limestone, and conglomerate. Great masses of diabase of uncertain age (p. 15) intrude the

[47] A. B. Parsons, *The Porphyry Coppers* (A.I.M. and M.E., 1923).
[48] M. J. Elsing and R. E. S. Heineman, *Arizona Metal Production* (Univ. of Ariz., Ariz. Bureau of Mines Bull. 140, 1936), p. 99.
[49] Description largely from Ransome, *op. cit.*

Apache and older rocks. Overlying the Apache group, without any recognizable unconformity to explain the apparent absence of the Ordovician and Silurian, is 325 feet of limestone, supposed to be Devonian. Conformably above the Devonian limestone is the light gray Carboniferous limestone, at least 1,000 feet thick.

After the deposition of the Carboniferous limestone the region was uplifted and eroded.

Cretaceous sediments were probably deposited, although no remnant of these is present in the region here particularly described. Their nearest known representatives are in the Deer Creek coal field, south of Gila River. The deposition of the supposedly Cretaceous beds was succeeded by andesitic eruptions, of which some of the products remain in the southern part of the Ray quadrangle.

The andesitic eruptions were followed by the successive intrusion of (1) quartz diorite, in small irregular masses and a few fairly large dikes; (2) granite, quartz monzonite porphyry, and granodiorite in masses, some of which, as the Schultze granite, are several miles in diameter; and (3) quartz diorite porphyry in dikes, sills, and small rotund bodies. The intrusion of the rocks of the second group was the cause of the original or hypogene metallization that, followed some time later by downward or supergene enrichment, gave rise to the disseminated copper ores of Ray and Miami. The time of the intrusion of the rocks in these three groups is not known but is thought to have been Laramide.

A period of active erosion, during which the coarse clastic material of the Whitetail conglomerate was washed by streams into local basins, followed the granitic intrusions, and this formation in turn was buried under a flow of dacite, probably in late Tertiary time. After this outburst the region was much faulted, vigorous erosion set in, and the generally coarse fluviatile deposit known as the Gila conglomerate was deposited, probably in late Tertiary time. This deposit has since been deformed by faulting and has been much dissected by the intermittent streams of the present drainage system.

STRUCTURE

The rocks of the area are cut by innumerable faults running in all directions. There is scarcely any flexing or folding of the beds. The structure is characterized by the dominance of deformation by faulting, mostly of the normal type.

The region east of Mineral Creek is marked by a network of block faults which have lowered the sedimentary rocks against pre-Cambrian schist. Roughly paralleling Mineral Creek and adjacent thereto is the Ray fault, striking northwest and forming the eastern boundary of ore enrichment. Branching from this fault and striking roughly east-west with a dip to the south is the North End fault, which, as its name implies, has limited enrichment on the north side of the ore body. Deposition on the west has been limited by the West End fault which strikes north-south.

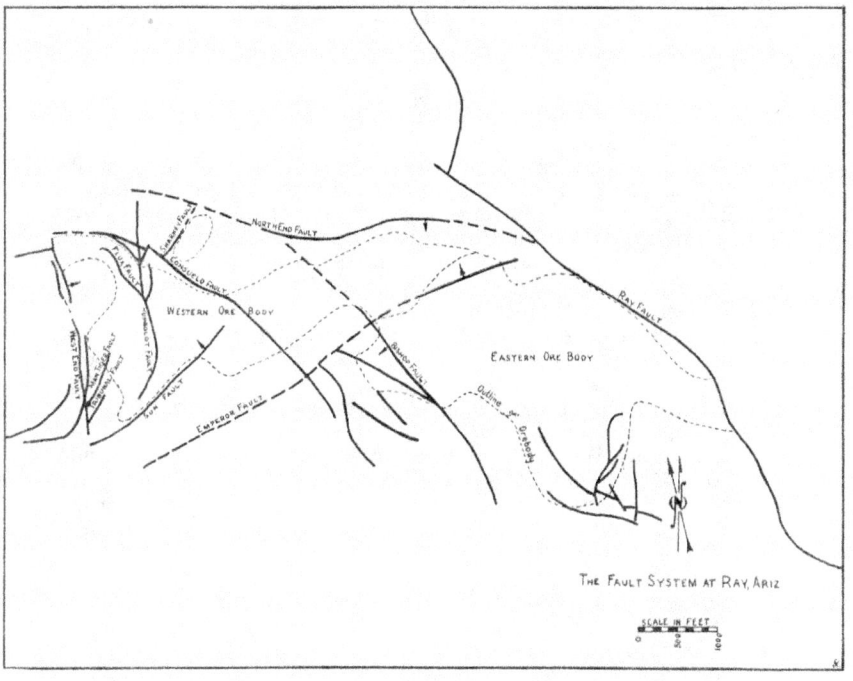

Figure 7.—Fault system at Ray. (By C. Leroy Hoyt.)

On the south the ore enrichment has been controlled by a series of diagonal faults (Fig. 7).

The fissuring that accompanied the early intrusion of the diabase into the schist established lines of weakness which were followed in a general way by the later porphyry intrusions and still later by the major faults, such as the Sun, Consuelo, and Emperor. The Ray, Bishop, Consuelo, and other faults strike approximately parallel to the general trend of the Dripping Spring Mountain Range. South of the Consuelo fault is a series of somewhat concentric circular faults generally dipping to the center and flattening out as they progressed downward, as shown by the Man Tiger, Tribunal, Sun, Flux, and the southern end of the Humboldt faults.

Ore Deposits

The ore deposit is a secondary enrichment of disseminated chalcocite, associated with and partially replacing primary pyrite in the pre-Cambrian Pinal schist and, to a slight extent, in Laramide porphyries. It is generally referred to as being a low-grade porphyry deposit. The ore body proper is a flat-lying mass, irregular in outline, and of variable thickness. The long axis extends roughly east and west for about 7,000 feet. It ranges in width from about 200 feet at the center to over 2,000 feet near the eastern and western extremities. The central constriction divides the ore into two sections which are called the "Eastern ore body" and the "Western ore body." The thickness of the ore as determined from drilling and development averages about 120 feet and ranges from 15 to more than 400 feet.

The area of oxidized capping is somewhat more extensive than that of the ore, but has the same general shape. Around the margin of the ore many of the drill holes pass directly from the oxidized capping into the unaltered primary protore. The thickness of the capping varies greatly, but its average is about 225 feet.

Doubtless as a result of the intrusion and solidification of the porphyries, the rocks were intricately broken by numerous small, irregular fissures which were permeable to the ore solutions.

Under the greater portion of the Eastern ore body a diabase sill that slopes gently to the east and north was more highly mineralized than was the surrounding schist. Chalcopyrite associated with the pyrite makes this diabase considerably higher in copper than the corresponding primary schist protore but is not of economic importance at the present time. This diabase sill is covered by a layer or blanket of gouge about 15 feet in average thickness, and it has acted as a dam to descending solutions from the oxide zone, causing them to deposit a large portion of their metallic burden in the gouge blanket. This enriched layer has been sufficiently high in copper to warrant some square-set mining in the past.

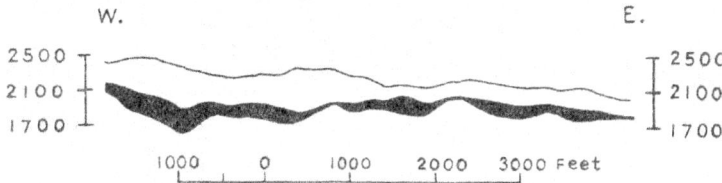

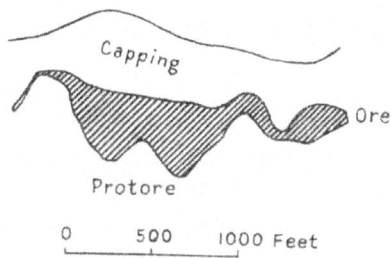

Figure 8.—Generalized sections through Ray ore bodies. (After Ransome, U.S.G.S. Prof. Paper 115.)

The concentration of the copper was greatly influenced by barriers that confined the descending solutions to well-defined areas.

One of the most important of these areas is in the triangular trough formed by the Consuelo, Sun, and West End faults. Near the center of this area in the vicinity of the Humboldt fault the greatest vertical thickness of the ore is attained. Along the West End fault, however, some sections have not received the enrichment found in the rest of this trough due to the fact that in these places erosion progressed more rapidly than oxidation.

North of the Consuelo and lying in the trough between the North End and Sharkey faults is a small, isolated ore body known as the Sharkey area.

A third important ore zone lies north of the Emperor fault in successive troughs formed by the intersections of the Emperor with the Consuelo, Bishop, North End, Ray, and other faults.

These three areas comprise the Western ore body. The Emperor fault controlled enrichment on the south side of the Western ore body for about 3,000 feet and formed a definite boundary between the Eastern and the Western ore bodies.

The Eastern ore body lies south of the Emperor in the section between the Bishop and the Ray faults and terminates in a series of unnamed faults at the southern end. Under a large part of this area the ore bottoms on the diabase previously mentioned.

From the preceding discussion the conclusion is drawn that the structural relationship of the fault system at the Ray property has been the controlling factor in the enrichment of the ore. During the process of enrichment the copper-bearing solutions which permeated the district were prevented from migrating laterally by fault barriers. The downward progress of the solutions was checked either by a flattening of the faults at depth, by troughs formed at their intersections, or, as in the Eastern ore body, by a diabase sill.

ACKNOWLEDGMENTS

Acknowledgments are due Dean LaGrange, Assistant Superintendent of Mines, and P. T. Whitehead, Geologist, Ray Mines Division, Kennecott Copper Corporation, for their assistance in the preparation of Hoyt's manuscript.

AJO DISTRICT[50]

BY JAMES GILLULY[51]

GEOGRAPHY

The Ajo copper district is in southern Arizona, in Pima County, about 43 miles south of Gila Bend, on the Southern Pacific Railroad, and 125 miles west of Tucson. It is in the extremely arid section of the state, with rainfall averaging less than 10 inches a year.

The district lies in the low desert plains, at altitudes ranging between 1,700 and 2,500 feet above the sea. The topography is hilly but not extremely rugged, with rather steep-sided hills rising abruptly above wide sloping pediments that merge into the alluvial intermontane plains.

HISTORY

Although the occurrence of copper at Ajo was established at least as early as 1750, it first came to the attention of English-speaking Americans in the days of the California gold rush of 1849. The first locations were made just after the Gadsden Purchase, but the early attempts at exploitation were unsuccessful owing to the low grade of the ore, the difficulty of water supply, and the extremely costly transportation. A renewed attempt to develop the deposits was made in 1894, but it too was ephemeral.

---

[50] Published by permission of the Director, U.S. Geological Survey. For a more detailed description of the district, see James Gilluly, *Geology and Ore Deposits of the Ajo Quadrangle, Arizona* (Univ. of Ariz., Ariz. Bureau of Mines Bull. 141, 1937). Paper prepared for the regional meeting of the A.I.M.&M.E. held at Tucson, Arizona, November 1-5, 1938.

[51] Geologist, U.S. Geological Survey.

At the beginning of the present century the brilliant success of the Utah Copper Company at Bingham inspired greater interest in the disseminated copper ores. Considerable promotion but little real development work was done at Ajo.

The active and successful development of the great deposit began in the fall of 1911 when the Calumet & Arizona Mining Co., of Bisbee, under the leadership of John C. Greenway, general manager, formed a reorganized New Cornelia Copper Company and began to test the property.

The successful exploitation of the deposit hinged upon development of a leaching process for the carbonate ores. The railroad from Gila Bend to Ajo was completed in 1915, and shipments of high-grade ores were made throughout 1916. An ample water supply was developed at the "water mine," about 600 feet deep, 7 miles north of Ajo. A 5,000-ton crushing, leaching, and electrolytic precipitation plant was completed by May, 1917. The oxidized ores were the sole source of copper until 1924 when, a 5,000-ton concentrator having been built, production began from the sulphides. In 1930 most of the oxidized ore having been exhausted, the leaching plant was closed. In the meantime, in 1928 and 1929, the sulphide concentrator was enlarged to a present capacity of 16,000 tons a day. Relatively minor changes would increase the capacity to 20,000 tons.

In 1929 the New Cornelia Copper Company was absorbed by the Calumet & Arizona Mining Co., and in 1931 this company was in turn consolidated with the Phelps Dodge Corporation.

GEOLOGY

The oldest rocks in the immediate vicinity of the Ajo ore body (Pl. XXI) are a series of lavas and associated tuffs, chiefly keratophyres and quartz keratophyres (locally called "rhyolites"), with subordinate andesite. It is possible that the keratophyric rocks were originally less sodic and owe their present chemical composition to later albitization.

The attitude of these volcanic rocks is uncertain, for although structures resembling flow phenomena are widespread they are of such diverse orientations, even on the same outcrop, as to preclude any confident deductions as to the original surfaces of the flows.

Into this volcanic series was intruded an elongated stock of porphyritic quartz monzonite with an external discontinuous shell of quartz diorite as much as 1,000 feet wide. The present outcrop of this intrusive mass is roughly wedge shaped, with the point in which practically all the known ore is concentrated projecting toward the southeast. The northern and northeastern limits are obscured by alluvium. As now exposed it is about 2 miles long and a mile wide at the north end. At the south end the monzonite has been found by diamond drilling to extend at least half a mile south of its surface limit in that direction.

The southeastern tip of the quartz monzonite (as exposed on

the surface) and the closely associated volcanic rocks are the host rocks of the ore deposit. The intrusive contact dips steeply south at the south end of the ore body (Pl. XXII), and it is reported[52] that "rhyolite" occurred on top of some of the hills, since removed in mining, so that it is likely that the roof of country rock has only recently been eroded from this part of the intrusive. Inasmuch as diamond drilling has demonstrated that the intrusive mass is relatively thin, with "rhyolite" beneath, it may be that the body as a whole is tabular, in a general way parallel to this contact.

The tip of the monzonite and the adjoining volcanic rocks are truncated at the south by a clean-cut, rather even erosion surface on which rests a great thickness of fanglomerate with intercalated volcanic rocks, now dipping 60 degrees or so to the south. The fanglomerate contains boulders of mineralized monzonite, diorite, and numerous rocks of exotic source, such as schist, coarse granite gneiss, and fossiliferous Carboniferous limestone, none of which are known to be of local derivation.

ORE BODY

The ore body is almost wholly in the monzonite, although volcanic rocks just to the southwest and southeast are also considerably mineralized. It is crudely elliptical in shape, about 3,600 feet long by 2,500 feet across. Its average thickness[53] is 425 feet, and the maximum about 1,000 feet. Most of the ore is in a rather flat lens, with a deeper northwestward-trending keel, but at the south end a tongue dips steeply southward to great depths.

The primary ore consists chiefly of chalcopyrite, with bornite and a little pyrite. The gangue is quartz and orthoclase, and the sulphides are distributed both in veinlets and in discrete grains through the altered monzonite. A little tennantite, considerable magnetite and specularite, and a little sphalerite and molybdenite also accompany the ore. The rock is highly orthoclasized, really pegmatized, along two main north-northwesterly zones, and the richest ore accompanies this more intensely altered rock. Chlorite and sericite after plagioclase are widespread, but the rock contains so much orthoclase and quartz that it is extremely hard and resembles fresh rock physically, in marked contrast with the soft, chalky-appearing ores of most of the "porphyry coppers."

The ore body was oxidized to a surprisingly level plane near the present water table, at an altitude of about 1,800 feet. There were local variations of as much as 50 feet, but for the most part the transition from sulphide to the oxidized zone was about as sharp as could be mined by steam shovel and was at nearly the

---

[52] I. B. Joralemon, "The Ajo Copper-mining District," *Am. Inst. Min. & Met. Eng. Trans.*, XLIX (1915), 593-609.

[53] G. R. Ingham and A. T. Barr, *Mining Methods and Costs at New Cornelia Branch of Phelps Dodge Corporation, Ajo, Arizona* (U.S. Bur. Mines Inf. Circ. 6666, 1932).

same altitude beneath hill and valley. The depth of oxidized ore thus ranged from 20 to 190 feet, with an average of about 55 feet.[54] The minerals of the oxidized ore were malachite with a little azurite, cuprite, tenorite, chrysocolla, hematite, and limonite. A little chalcocite occurs close beneath the bottom of the oxidized zone.

The fact that in most of the ore body the tenor of ore was essentially the same in oxidized and subjacent sulphide ore seems to show, in connection with the rather insignificant quantity of chalcocite, that there was little migration of copper during weathering but that the sulphides were oxidized in place. In this respect the Ajo ore body differs from all the other great disseminated deposits of the Southwest, in each of which supergene chalcocite enrichment was essential to the production of commercial ore.

The supergene chalcocite of the Ajo deposit is found in insignificant amounts over most of the area of the pit at about the original ground-water level, but at the south end of the ore body there is a crescentic outcrop of chalcocitic ore. This zone pitches 60 degrees S., parallel to the dip of the overlying fanglomerate, and has been followed by the diamond drill to depths of more than 200 feet below sea level. This chalcocite zone is overlain by a reddish weathered zone containing cuprite, native copper, and hematite and obviously represents an old supergenely enriched zone formed prior to the deformation of the fanglomerate and probably prior to its deposition. The oxidized and enriched zones have been deformed in this region and dislocated by faults of several hundred feet displacment, as shown by diamond drilling.

The existence of this zone, in which ore containing 3 to 4 per cent of copper is common, indicates that the absence of enrichment in the present cycle of erosion may not be due, as has been commonly suggested, to the small amount of pyrite in the ore. It is possible that climatic factors may be more important in this connection, although it is possible that the pyrite factor may be dominant after all—long-continued erosion, depression of the water table, and consequent weathering of much larger quantities of pyrite than were available during the present cycle being necessary to produce the older chalcocite enrichment.

The close association of the ore deposition with the pegmatization of the quartz monzonite porphyry in conjunction with the content of magnetite and specular hematite is sufficiently indicative of its igneous origin. The emplacement was largely effected by metasomatic processes, guided by the widespread fissuring along dominantly north to northwest lines.

The regional geologic setting of the Ajo deposit is not yet well enough known to furnish clear-cut data as to its age. No fossiliferous rocks are involved in the near-by structures. The present erosion surface has been masked by volcanic flows, now largely removed. This surface is carved across the steeply tilted fan-

---

[54] Ingham and Barr, *op. cit.*

glomerate, and that in turn rests on a presumably deeply eroded surface of mineralized monzonite. Thus if the deposit is Tertiary, as has been suggested, its formation has been succeeded by a long and rather involved history. On the other hand, intense contact metamorphism of Carboniferous (?) limestones in the Growler Mountains to the south may indicate that the major intrusives were post-Carboniferous. As far as direct data go, then, nothing can yet be said with certainty as to its age—the probabilities lie between Permian and early Tertiary.

Mining Methods

The deposit is mined by the open-cut method, with power shovels operating on benches at vertical intervals of 30 feet. The present outline of the pit and its relation to the ore body are shown in Plate XXII. Inasmuch as the oxidized part of the ore body was practically as productive as the sulphide part, there was here no stripping problem of the sort confronting most of the disseminated deposits of the Southwest. To January, 1931, less than 7,000,000 tons of waste had been moved in the mining of 32,400,000 tons of ore, a ratio of 0.21 ton of waste to 1 ton of ore. Much of this waste occurred within the ore body and was not overburden. As the depth of the pit increases, however, a larger proportion of waste will have to be moved in order to maintain a safe angle of slope.

Magma Mine Area, Superior[55]

By

M. N. Short[56] and Eldred D. Wilson

Situation

The Magma Mine is at Superior, in the Pioneer district, 15 miles southwest of Miami and 12 miles northwest of Ray. Superior lies at an altitude of 2,850 feet on the eastern margin of a small basin-shaped valley in the mountainous region between the Superstition and Pinal ranges.

History

The Magma vein was located in 1874 or 1875 during the period of exploration that led to the discovery of the deposits at Silver King, 2 miles farther north, and at Globe. A vertical shaft, the Silver Queen, was sunk 400 feet, and a few pockets of silver-enriched chalcocite were discovered. Activity largely ceased by 1893. In 1910 the Silver Queen Mine was optioned by William Boyce Thompson and associates who organized the Magma Copper

[55] Paper prepared for the regional meeting of the A.I.M.&M.E. held at Tucson, Arizona, November 1-5, 1938.
[56] Professor of Petrography, University of Arizona.

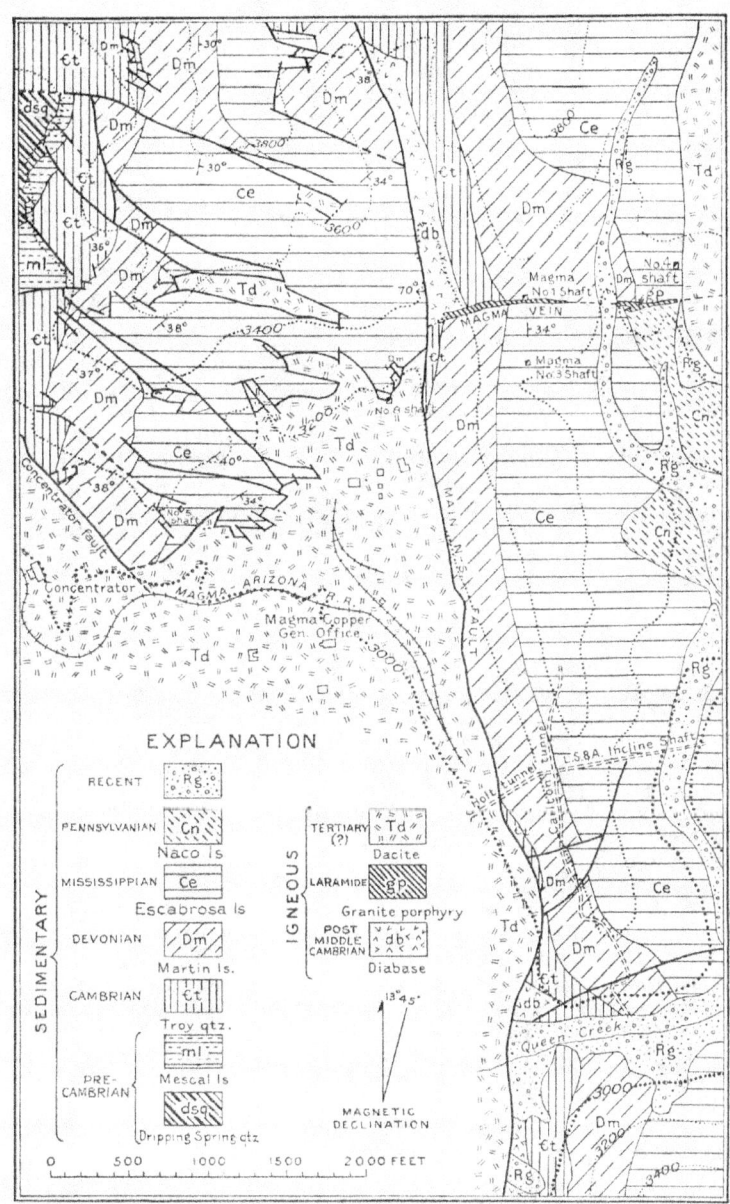

Figure 9.—Geologic surface map of vicinity of Magma Mine. (By Short and Ettlinger.)

Company. This company has maintained nearly continuous operations to the present time. It built a railroad and a concentrator in 1914 and completed its smelter in 1924. The mine, which has reached the 4,000-foot level, is now the deepest in Arizona. The constant high tenor of its ore has made the Magma one of the low-cost producers of the United States, despite depth, temperature, and heavy ground.

Production of the Magma Mine from 1914 to 1937, inclusive, has been as follows:

Ore (tons)	Copper (lbs.)	Gold (oz.)	Silver (oz.)	Total value
3,984,414	468,149,862	144,518	12,331,681	$72,412,678

After about 1902 the Lake Superior and Arizona Mining Company drove the Carlton and Holt tunnels (Fig. 9) and sank a 1,400-foot incline in ground between the Magma vein and Queen Creek. Copper ore, containing some silver and gold, was produced in 1907 and during the World War. Since 1920 the ground has been owned by the Magma Copper Company. In 1932 T. D. Herron and C. Laster leased the mine and opened large bodies of gold ore. Their production during 1932 to 1937, inclusive, as stated in the annual reports of the Magma Copper Company, amounted to 56,649 tons of ore which yielded 372,420 pounds of copper, 53,162 ounces of silver, and 31,598 ounces of gold.

Rocks

At the Magma Mine the rocks are essentially the same as at Ray (p. 81 and Pl. III). The oldest formation, the Pinal schist, of early pre-Cambrian age, has been penetrated by the 3,600-foot level of the Magma Mine. There it is unconformably overlain by the late pre-Cambrian Apache group which normally consists of the basal Scanlan conglomerate, 15 feet thick, overlain by 150 feet of Pioneer shale, 15 feet of Barnes conglomerate, 450 feet of Dripping Spring quartzite, and 225 feet of Mescal limestone, in places with a basalt flow at the top. In the bluff east of Superior the Apache group is overlain by 400 feet of Middle Cambrian Troy quartzite, succeeded by 340 feet of Devonian Martin limestone, 175 feet of Mississippian Escabrosa limestone, and 800 feet of Pennsylvanian Naco limestone.

The Apache and older rocks are intruded by large masses of diabase. In the Magma Mine it forms two sills of which the lower, 1,120 feet thick, invades Pioneer shale, and the upper, 2,000 feet thick, lies between the Troy and Dripping Spring quartzites. Apparently the upper sill engulfed the Mescal limestone, which has not been found in the mine. The age of the diabase has been regarded as early Mesozoic (?) by Ransome, late pre-Cambrian by Darton, and post-Middle Cambrian by Short (p. 15).

The Naco limestone and underlying formations are invaded by dikes and sills of quartz monzonite porphyry. One of these dikes is followed in the upper levels of the mine by the Magma fault and Magma vein. Several other dikes of this rock have been found by diamond drilling. The porphyry is believed to have been intruded during Laramide (late Mesozoic or early Tertiary) time, as were the Schultze granite at Miami, the granite porphyry at Ray, and the quartz diorite at Silver King. All of these intrusives are probably offshoots of the central Arizona batholith[57] with which the mineralization at Superior, Miami, Globe, Ray, and Silver King is believed to be genetically connected.

Dacite flows, tuff, and agglomerate cover much of the region. This volcanic material is more than 1,000 feet thick on Apache Leap, a short distance east of Superior; 1,300 feet thick, in Picketpost Mountain, 3 miles west of Superior; and 2,500 feet thick in the Superstition Mountains. It was erupted during early Tertiary time, but long after the mineralization, upon a deeply eroded surface of the tilted sedimentary beds. In Apache Leap the hollows of this old erosion surface show remnants of a predacite conglomerate equivalent to the Whitetail conglomerate that, near Ray, attains a thickness of more than 800 feet. Most of the oxidation of the ore bodies is believed to have taken place during the long period of erosion in which this conglomerate accumulated.

Later than the dacite and the Main fault (p. 94) are a few small dikes and pluglike masses of amygdaloidal basalt.

Gravel, sand, and silt, formed by late Tertiary and Quaternary erosion and sedimentation, mantle the valley floor west of Superior.

STRUCTURE

During the Laramide revolution, in late Mesozoic or early Cenozoic time, the region underwent extensive deformation, accompanied or closely followed by intrusion of the quartz monzonite porphyry. Due to this deformation the Apache and Paleozoic beds strike north-northwestward, dip about 30 to 35 degrees eastward, and are broken by fractures and faults of eastward trend and by bedding or strike faults. Some of the eastward-trending fissures were sufficiently deep seated to be occupied by dikes of the quartz monzonite porphyry. Reopened by further movement, they afforded permeable channels for mineralizing solutions. The Magma fault, which is the locus of the Magma vein, represents this type of fissure. The Lake Superior and Arizona vein is within the zone of a bedding or strike fault.

Magma fault.—The Magma fault strikes about S. 80 degrees W. Its dip averages about 70 degrees N. from the surface to the 800-foot level, vertical from the 800- to the 900-foot level, and about 80 degrees S. from the 900- to the 4,000-foot level (Fig. 10). Its reversal of dip occurs where the dominant wall rocks change from sedimentary beds to diabase. This fault is not a single dislocation but a zone of closely spaced fractures 5 to 40 feet wide with well-defined footwall and hanging wall. On the west it is dislocated

---

[57] M. N. Short and I. A. Ettlinger, *Ore Deposition and Enrichment at the Magma Mine, Superior, Arizona* (Am. Inst. Min. Eng., Trans.), LXXIV (1927), 183.

by later transverse faults, but eastward it extends more than 8,000 feet without much dislocation.

The south side of the Magma fault has been dropped 500 feet relative to the north side. Due to their eastward dip, the beds on the north side of the fault show a relative horizontal displacement of 400 or 450 feet eastward. The real horizontal displacement exceeds the vertical.

Main and Concentrator faults.—Much later than the Magma fault and later than the dacite are the Main and Concentrator normal faults, between which is a mosaic of fault blocks (Fig. 9).

The Main fault strikes northward and dips steeply westward above the 2,800-foot level, below which its dip flattens to about 35 degrees W. Its vertical displacement apparently amounts to about 500 feet in the mine workings and increases southward, and its west or hanging-wall side has shifted the Magma vein 1,400 feet southward.

The Concentrator fault strikes S. 80 degrees E., dips steeply southward, and locally forms a crushed zone more than 10 feet wide. Its apparent stratigraphic throw is at least 3,500 feet.

Ore Bodies

Distribution.—Exploration of the Magma vein for a length of 8,700 feet and a depth of 4,000 feet has revealed three groups of ore shoots. Most of the production has come from the main or middle ore body. The west ore body was west of the Main fault, below the 2,250-foot level. It consisted of copper ore, now largely mined out. The east ore bodies lie between the main crosscuts and No. 6 shaft. They consist largely of zinc-copper ore of which comparatively little has been mined.

The gold-bearing zones in the Lake Superior and Arizona Mine have been opened north of Queen Creek for a length of more than 3,000 feet (Fig. 9), within which seven principal ore shoots have been found.

Magma Vein

Outcrop.—Considering the size of the Magma ore bodies, the outcrop of its vein is inconspicuous. Above the main ore body the bleached, faulted porphyry dike is stained by copper and iron and locally contains small masses of residual chalcocite.

Main ore body.—The main ore body has its apex between the 400- and 500-foot levels and extends to the lowest workings of the mine. Near the 1,500-foot level it is joined on the west by a branch (Pl. XXIII) that forms an apex a short distance above the 1,200-foot level. In places it has been stoped for a length of 1,200 feet with widths ranging from less than 5 to more than 30 feet. The axis pitches westward, essentially at right angles to the beds.

The ore shoot is of replacement type. Its gangue consists of crushed, altered wall rock and a large proportion of introduced silica. Small stringers of ore minerals occur in the walls, but commercial ore is confined to the fault zone itself.

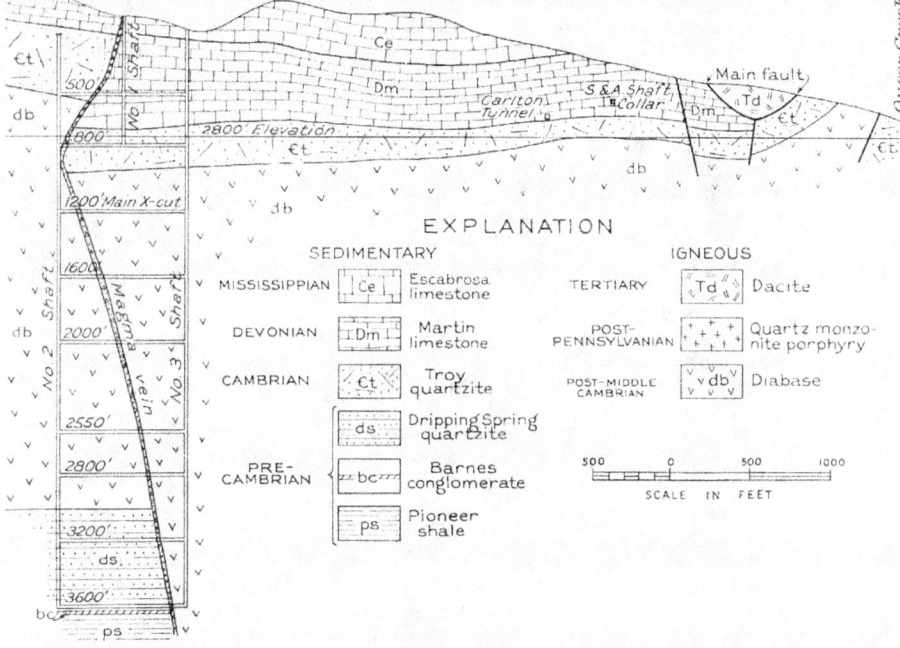

EXPLANATION

SEDIMENTARY

MISSISSIPPIAN [Ce] Escabrosa limestone

DEVONIAN [lDm] Martin limestone

CAMBRIAN [Ct] Troy quartzite

[ds] Dripping Spring quartzite

PRE-CAMBRIAN { [bc] Barnes conglomerate

[ps] Pioneer shale

IGNEOUS

TERTIARY [Td] Dacite

POST-PENNSYLVANIAN [+ + +] Quartz monzonite porphyry

POST-MIDDLE CAMBRIAN [v db v] Diabase

SCALE IN FEET

Figure 10.—North-south geologic sections through shafts at Magma Mine, Pinal County.

There are no replacement ore bodies in the limestones. The ore shoot is richest where its walls are of diabase. The total tonnage of sulphide ore in that part of the ore shoot where quartzite or shale forms both walls of the vein is perhaps equal to that developed where diabase forms either or both walls but is of lower grade. For example, on the 3,000-foot level where the walls are mainly sedimentary rocks, the ore consists of low-grade pyrite together with some streaks and admixed masses of chalcopyrite but with very little bornite. On the 3,600-foot level where the walls are the lower diabase sill, the ore consists mainly of bornite, together with chalcopyrite, tennantite, enargite, and chalcocite.

On most levels, bornite is the principal hypogene mineral, although in places chalcopyrite predominates. Pyrite is abundant with the chalcopyrite but less so where bornite is the dominant mineral.

Above the 1,200-foot level, the western branch of the main ore shoot contains no copper minerals but consists of sphalerite and galena. Below that level it changes abruptly into a bornite-rich ore with little or no zinc and lead. There is also a marked tendency for zinc and lead to occur on the eastern, but not on the western, margin of the ore shoot.

Tennantite is important from the 1,200- to the 3,600-foot level where enargite takes its place in the western part of the ore body.

Deep-level chalcocite accounts for about 5 per cent of the copper in the lower levels of the main ore body. There it invariably occurs intergrown with bornite and not alone or intergrown with other minerals. The weight of evidence favors its hypogene origin, as discussed in an earlier publication.[58] A comparison of specimens collected indicates chalcocite to be approximately of the same abundance on the 2,000- and 3,600-foot levels.

In the main ore body supergene chalcocite persisted to about the 800-foot level. The bottom of the oxidized zone dips eastward and corresponds rather closely with the base of the Martin limestone.

West (No. 5) ore body.—The sinking of No. 5 shaft between the Main and Concentrator faults disclosed an ore body in the Magma vein that extended from 100 feet above the 2,250-foot level to approximately 100 feet below the 2,500-foot level. Stoped for a length of approximately 250 feet and a width of 15 feet, this ore body averaged 7 per cent copper, principally as bornite. The ore was strongly oxidized, with chalcocite, cuprite, and native copper locally abundant on the 2,550-foot level. Oxidation is relatively more intense near the Main fault, which indicates part, at least, of the oxidation to be later than the fault. Much of it, however, may antedate the downfaulting of the ore body.

East ore bodies.—Several small ore shoots have been found in the Magma vein east of the main crosscuts between the 1,600- and 3,200-foot levels (Pl. XXIII). Their ore is principally sphalerite,

---

[58] Short and Ettlinger, *op. cit.*, p. 207.

but copper minerals are locally abundant. In places, strong oxidation extends below the 2,000-foot level.

These ore bodies, which have not been extensively mined, form a valuable reserve of zinc and copper ores.

Wall-rock alteration.—Within the Magma fault zone the diabase is bleached, with a development of quartz, sericite, and carbonates. As this rock had already been rendered impervious by extensive serpentinization and uralitization, the influence of the ore solutions dies out a short distance from the permeable fault zone.

Alteration of the siliceous and aluminous sedimentary beds likewise consists of silicification and sericitization, but less intense than in the diabase.

GOLD DEPOSITS

The rocks in the vicinity of the Lake Superior and Arizona Mine are Cambrian quartzite and Devonian limestone which strike northward and dip about 30 degrees E. The vein occurs within the zone of a strike fault that has brecciated the quartzite-limestone contact and the lower beds of the limestone. As stated by Ransome,[59] this brecciation is associated in surface exposures with limonite, manganese oxide, quartz, and hematite, and in places with malachite and chrysocolla.

About 1902, this zone was opened by a shaft, some 1,400 feet deep, that inclines 26 degrees E. and connects with eight levels of drifts. Most of the drifting is on the second or Carlton tunnel level which extends southward for some 2,000 feet and opens into Queen Creek Canyon. East of its portal is an old vertical shaft, the Vivian, that taps the vein at a depth of 140 feet. These old workings, which were mainly in the footwall portion of the vein, exposed only a few small bodies of oxidized copper ore, and material that contained generally less than 0.2 ounce of gold per ton.

The present lessees, by crosscutting along transverse fissures for a few feet towards the hanging wall, discovered seven or more shoots of gold ore within a horizontal distance of 3,000 feet. These ore bodies average 4 feet wide by 15 feet long, and the most persistent one extends, with interruptions, to the bottom level of the mine. The ore consists mainly of hematite, limonite, and fine-grained grayish to greenish yellow quartz of epithermal aspect. As a rule, the gold is spongy to fine grained and erratically distributed. According to Mr. Herron, the ore mined contained generally an ounce or more of gold and an ounce of silver per ton.

BIBLIOGRAPHY

Ransome, F. L., Copper Deposits near Superior, Arizona, U.S. Geol. Survey Bull. 540, pp. 139-58, 1913.
Ransome, F. L., Copper Deposits of Ray and Miami, Arizona, U.S. Geol. Survey Prof. Paper 115, 1919.

[59] F. L. Ransome, *Copper Deposits near Superior, Arizona* (U.S. Geol. Survey Bull. 540, 1913), p. 155.

Browning, W. C., and Snow, F. W., Geology and Operations of the
   Magma Mine, Eng. & Min. Jour., Vol. 119, p. 197, 1925.
Short, M. N., and Ettlinger, I. A., Ore Deposition and Enrichment
   at the Magma Mine, Superior, Arizona, Am. Inst. Min. Eng.,
   Trans., Vol. 74, pp. 174-222, 1927.
Ettlinger, I. A., Ore Deposits Support Hypothesis of a Central
   Arizona Batholith, Am. Inst. Min. Eng., Tech. Pub. 63, 1928.
Ettlinger, I. A., and Short, M. N., The Magma Mine, Superior, 16
   Int. Geol. Cong., Copper Resources of the World, Vol. 1, pp.
   207-13, 1935.
Wilson, Eldred D., Arizona Lode Gold Mines, Univ. of Ariz., Ariz.
   Bureau of Mines Bull. 137, pp. 168-70, 1934.
Gardner, E. D., Mining Methods and Costs at Herron and Laster
   Lease, Superior, Arizona, U.S. Bureau of Mines Inf. Circ. 6799,
   1934.

Bagdad Mine, Eureka District[60]

By

B. S. Butler and Eldred D. Wilson

Introduction

The Bagdad Mine is in the Eureka district, western Yavapai
County, 27 miles by road from Hillside, a station on the Santa Fe
railway. Bagdad camp is on Copper Creek, a few miles upstream
from its junction with Burro Creek, at an altitude of about 3,200
feet.

Little has been published regarding the geology of the Bagdad
area, though several studies have been made and reports prepared for the Bagdad Copper Corporation and its predecessors.

The following discussion is largely based on these reports and
other data furnished by the company. The map (Pl. XXV) was
originally prepared by Rogers, Mayer, and Ball in 1918 and modified by H. N. Witt and P. C. Benedict in 1926.

The writers are particularly indebted to J. W. Still, mine
manager, for information.

History and Production

Prior to 1907, eight claims in the Bagdad area were patented
by John Lawler. Between 1907 and 1910 this group was obtained
by the Copper Creek Development Company, Inc., which did
additional development work and located additional claims which
were later patented. By various successions the property was
acquired in turn by the Arizona Nevada Copper Company, by the
Bagdad Copper Company, by the Arizona Bagdad Copper Company, and in March, 1927, by the Bagdad Copper Corporation.

---

[60] Paper prepared for the regional meeting of the A.I.M.&M.E. held at
   Tucson, Arizona, November 1-5, 1938.

Additional development work was done by each of these companies. In the past considerable underground work has been done for the purposes of checking churn drill assays, to provide ore for metallurgical test work in pilot mills, and to check the caving system of mining proposed for a larger operation. The estimated ore reserves, hereinafter mentioned, are based on information obtained from 123 churn drill holes, approximately 20,000 feet of underground work, and approximately 8,000 feet of underground diamond drilling.

The property now has mill and power plant equipment installed sufficient to handle 250 tons per day.

During 1937 a total of 75,512 tons of ore with an average copper content of 1.37 per cent yielded 1,792.76 tons of concentrates averaging 42.87 per cent copper.

Operating costs of delivering concentrates on cars at Hillside, Arizona, were $2.045 per ton, distributed as follows:

Mining	$0.732 per ton
Milling	1.021 per ton
Camp, etc.	0.292 per ton

Production in 1937 was 1,537,396 pounds of copper at a cost of 0.137 cents per pound.

Rocks

The oldest rocks at Bagdad are amphibolite and mica schists which, because of their lithology and metamorphism, have been correlated with the pre-Cambrian Yavapai schist of the Prescott-Jerome region. Their composition suggests derivation from both igneous and sedimentary rocks. Intrusive into the schists is a granitic rock that resembles the pre-Cambrian Bradshaw granite. The presence of schist suggests that the Bagdad area is near the margin or in the roof of the granite batholith, which is of wide extent. Both the granite and the schist contain abundant pegmatite bodies, also suggestive of the margin or roof of a batholith. The rock in which the Bagdad ore occurs is intrusive into the schist and pre-Cambrian granite.

Dr. C. P. Berkey, in a report to the company, classifies the later intrusive as granite porphyry that ranges widely in composition. Locally it is known as "monzonite." Both orthoclase and plagioclase are present with abundant quartz. Biotite is variable in amount though abundant in much of the rock. The age of this intrusion is not known; it may be Laramide.

Overlying the pre-Cambrian rocks and the later intrusive body is an irregular thickness of sedimentary material laid down on an old land surface. This material is largely conglomerate that fills old valleys. Locally it has some marly lake beds near the top. Capping the surrounding mesas and covering the conglomerate and all but the higher points of the earlier rocks are flows of basaltic lavas. The sedimentary material and the lava are presumably of late Tertiary or Quaternary age.

Plate XXIV.—Bagdad camp, looking up Copper Creek. Granite porphyry in foreground; mesa capped by basalt and underlain by conglomerate in background.

Recent erosion has cut steep-sided canyons into the mesa areas and exposed the prebasaltic rocks. Active dissection is still in progress. The present drainage lines differ from the earlier drainages and in the vicinity of Bagdad have not cut to the depth of the earlier dissection.

STRUCTURE

Little is known of the pre-Cambrian structure of the Bagdad area. Since a large part of the area and also part of the outcrop of the later intrusive rock is covered by basalt, it is not evident what structures may have controlled the location of the later intrusive body.

Following the later intrusive activity the rock was highly fractured, with some movement on the stronger breaks.

In most exposures the fractures show many directions and attitudes that are not readily reduced to definite systems. Some of the fractures, however, are strong and persistent and have been followed for long distances by mine workings. The general strike of these stronger mineralized fractures is northwest. The Black Mesa breccia area has an elongation in an east-northeast direction but without obvious strong fractures. Later than the mineralized fissures and apparently later than the gravels and the basalt are northeasterly striking faults of small displacement that offset the mineralized fissures. One of the strongest faults recognized is the Hawkeye which trends northwestward with the general direction of the mineralized fissures. As indicated on Plate XXV, it is west of the main developed area.

Plate **XXVI.**—Near view of shattered granite porphyry.

MINERALIZATION

A conspicuous feature around Bagdad is the red and brown iron stain on the rocks, particularly all exposures of the granite porphyry. Much of it is highly colored. This extensive staining indicates a rather widespread mineralization.

Prospecting in the mineralized granite porphyry has been mainly of two types, first, of the more prominent fissures, and, second, of disseminated deposits.

Development has been largely in the porphyry east of the junction of Copper and Marooney creeks, between the streams and the basalt-covered mesa, though a few drill holes have been sunk through the basalt and conglomerates of the mesa. The prospected area is also largely east of the Hawkeye fault.

Recent operations have been in the eastern part of the Bagdad area on both sides of Copper Creek where it flows southward. Here, the granite porphyry is broken into fragments of different sizes by fissures that strike and dip in various directions. As already noted, two directions are prominent—northwesterly mineralized fissures and northeasterly faults that offset the mineralized fissures. The spacing of premineral fissures in the developed area shows considerable irregularity. In some parts a hand specimen contains several fissures, whereas in other parts the rock is distinctly blocky with its prominent fractures spaced several inches or more apart.

The stronger fissures contain well-defined veins that range from a fraction of an inch to more than a foot in width. In general, they have not proved large and rich enough to mine as simple veins.

The smaller fractures are also mineralized. Some sulphide grains seem to be disseminated in unfractured rock, but they are probably connected with inconspicuous fissures.

The degree of mineralization appears to be related to the closeness of fractures. Coarse, blocky ground appears less mineralized than the more closely fractured areas.

The alteration of the rock is less intense than in most disseminated deposits. Much of the feldspar is clouded with sericite, and the biotite is in part altered to chlorite or bleached to muscovite, but there has been no complete working over of the minerals, and the rock generally looks fresh. Along the strongly mineralized fissures, however, the rock is more strongly sericitized. Like most of the copper deposits of the Southwest, the Bagdad deposit can be separated into three zones—namely, the oxidized zone, the zone of sulphide enrichment, and the primary lean sulphide zone.

The alteration of the deposit occurred in large part during the period in which the preconglomerate topography developed, though it probably continued during the accumulation of the conglomerate. It has certainly been active in the present cycle of erosion, which has produced the steep-sided valleys.

The present surface, especially on the higher slopes, is rather free from copper stains and appears to be rather completely leached of any copper that may have been present.

The cross sections of the ore body (Pl. XXVII) give a rather representative picture of the distribution of copper. Probably the near-surface cappings were not assayed where low in copper. Generally the amount of copper in the oxidized zone increases with depth, and in places just above the secondary sulphide zone it may approach the copper content of the sulphide zone. Ordinarily, however, it is distinctly of lower grade than the sulphide zone, and probably no large bodies of it would exceed 0.5 per cent copper. In total, however, a very considerable amount of the copper is in the oxidized zone.

The secondary sulphide zone consists of veinlets of pyrite and chalcopyrite partially replaced by chalcocite. The copper content of the enriched sulphide zone in general is highest just below the oxide zone and decreases gradually towards the primary zone. In the upper, richer portion of the secondary sulphide zone the average copper content is probably three to four times that in the primary zone, indicating a very considerable movement and enrichment of copper.

Some of the prominent fissures locally have the primary sulphide largely replaced by chalcocite, resulting in high-grade specimen ore.

The thickness of the enriched sulphide zone ranges considerably (Pl. XXVII). The lower boundary is, of course, drawn at what is regarded as the limit of profitable ore and not necessarily or likely at the boundary between secondary and primary ore.

As in many deposits, the primary sulphide zone beneath the enriched zone has not been extensively prospected or developed. Pyrite and chalcopyrite are the sulphides present, and the copper content in general does not appear to exceed 0.5 per cent.

Notable amounts of molybdenite occur in thin, widely scattered quartz veins that generally dip at low angles southeastward. These occurrences seem to account for the small amount of molybdenum in the concentrates. Molybdenite was not noted in the copper veins, though it is possibly present.

Some prospecting has been carried on west of the main developed area of disseminated copper (Pl. XXV).

The Black Mesa deposit consists of a brecciated area with elongation in a general east-northeast direction (Pl. XXV). This breccia is probably due to faulting. On the surface it is cemented with quartz and limonite. A tunnel driven into the breccia has encountered sulphides. So far as shown by this limited development, the sulphides are most abundant on the margins of the breccia.

The Giroux and Paul tunnels have prospected an area between the Black Mesa and the main Bagdad areas.

QUANTITY AND GRADE OF ORE

The following figures of quantity and grade of ore have been taken from the prospectus of the Bagdad Copper Corporation, May 11, 1938.

Ore estimates, made in various reports to the company and its predecessors, range from 5,000,000 to 35,000,000 tons, of which the estimated copper content ranges from 1 per cent to 1.93 per cent.

The latest report on the property, by Whitaker and Schlereth, of Denver, Colorado, dated June, 1937, estimated 6,000,000 tons with an average of 1.47 per cent copper. No estimate was made of oxidized ore.

The main deposit was developed by churn drill, and a part of the area has been opened for mining by the undercut caving system.

The capacity of the mill in 1937 was approximately 250 tons per day. The Whitaker and Schlereth report recommended a 1,000-ton mill as the most efficient for the property.

Structural Control of the Ore Deposits at Tombstone, Arizona[61]

By

B. S. Butler and Eldred D. Wilson

Introduction

Tombstone, Arizona, from its discovery by Ed. Schieffelin in 1877 through the eighties, was the best-known mining camp in the Southwest. During this period it produced the larger part of the approximately $37,000,000 worth of metals that it had yielded. Since those boom days it has been a consistent producer in a small way and during 1933-36 yielded more than a million dollars' worth of metals.

Geologic History

Rocks ranging in age from early pre-Cambrian to present are exposed at Tombstone. The geologic section (Pl. III) shows the general character of the sedimentary rocks, and the map (Pl. XXVIII) the distribution of the rocks.

The exposure of pre-Cambrian rock is too limited to reveal much of its history, though the schist indicates the long period of sedimentation and deformation common to the early pre-Cambrian of the Southwest.

Between early pre-Cambrian and Paleozoic the area was deeply eroded, and on this erosion surface of little relief was deposited a thick series of Paleozoic sediments, prevailingly limestone. During long periods, notably the Ordovician, Silurian, and early Devonian, sediments were not deposited, but little or no angular discordance appears in the Paleozoic section.

From Paleozoic to Mesozoic the sedimentation changed from prevailingly limestone to prevailingly sandstone and shale with subordinate limestone. No pronounced angular discordance marks this change, although, as Ransome has shown, faulting and extensive erosion occurred between the deposition of late Paleozoic and Mesozoic rocks in the neighboring Bisbee district. Pebbles of volcanic rocks in the Mesozoic beds at Tombstone indicate igneous activity in the interval between the deposition of the two series of sedimentary rocks, but igneous rocks of this age have not been recognized within the Tombstone area.

The deposition of the Mesozoic sedimentary rocks was followed by a period of deformation and igneous activity, the age of which is not closely known but which may be essentially the same as that of the Laramide revolution in the Rocky Mountain region. The ore deposits are associated with the folding, faulting, igneous intrusions, and fissuring of this period.

---

[61] For a more detailed description, with numerous structure sections, see B. S. Butler, E. D. Wilson, and C. A. Rasor, *Geology and Ore Deposits of the Tombstone District, Arizona* (Univ. of Ariz., Ariz. Bureau of Mines Bull. 143, 1938).
Paper prepared for the regional meeting of the A.I.M.&M.E. held at Tucson, Arizona, November 1-5, 1938.

Both faulting and igneous activity also occurred, in a mild degree, after deposition of the Cenozoic sediments that occupy the valley floors.

One of the early structures is the uplifted, eastward-tilted Ajax Hill block which is bounded on the north by the Prompter fault and on the south and west by faults. The west boundary fault, which brings Mesozoic rocks against pre-Cambrian, represents a throw greater than the total thickness of the Paleozoic rocks. The north and probably the south boundary faults decrease in throw eastward from the junction with the west boundary fault. The eastward tilting is confined to the western part of the block. Within the block are open east-west folds.

The east-west folds within the Ajax block, the generally southeast-northwest trending Tombstone basin north of the Ajax block, and the folds within that basin were possibly formed at nearly the same time as the Ajax block structure. Following and possibly overlapping the folding and faulting were north-south fissures of steep westerly dip and small displacement. The west boundary fault of the Ajax Hill block, however, with large displacement, has this attitude. The north-south fissuring was followed by igneous activity both extrusive and intrusive. Southwest of the district an area of volcanic breccia and flows is intruded by quartz latite porphyry that, within the Tombstone area, intrudes the Mesozoic sedimentary rocks. Similar porphyry occupies some of the north-south fault fissures that cut both the Paleozoic and Mesozoic sedimentary rocks. In the northwestern part of the district a mass of granodiorite, probably intruded along a fault, separates the Mesozoic and Paleozoic rocks. Dikes of similar composition are present in the sedimentary rocks.

The relative age of the main intrusive bodies is not definitely known. They were probably of one general period, but it seems likely that the granodiorite is later. Dikes intrude both the quartz latite porphyry and the granodiorite, and basaltic rocks are younger than the late sediments of the valley fill.

Fissures of generally northeastward strike and steep dip cut the pre-Tertiary sedimentary rocks, the earlier intrusive rocks, and the earlier structures. These northeasterly striking fissures are the main ore channels, and their intersection with favorable structures and favorable rocks determines the location of ore shoots.

Later than the ore fissures are faults with north to northeasterly strike, represented by the Tranquility fault, and faults with northwesterly strike, represented by the Grand Central fault.

SUMMARY OF STRUCTURAL EVENTS

1. Generally southeast-northwest and east-west folds and faults.
2. North-south faults and fissures (dike fissures).
3. Intrusion of quartz latite porphyry, granodiorite, and allied rocks.

4. Northeast-southwest fissures.
5. North to northeast and northwest faults.

INFLUENCE OF ROCK CHARACTER ON THE STRUCTURE

The different response of the massive Paleozoic limestone series and the weak Mesozoic shaly series to stresses is well shown in Tombstone basin, an asymmetrical synclinal fold with easterly to southeasterly pitch, which occupies the northern, most productive part of the district. The rocks along the southern border of the basin dip steeply, and near the Prompter fault are slightly overturned. Within the basin are open secondary anticlines and synclines on the larger Tombstone basin syncline.

Folds of a third order occur as corrugations on the secondary anticlines and synclines. The third-order anticlines are locally termed "rolls," and on them are located many ore shoots. The "rolls" are better developed on anticlines than on synclines, and better on close, steep anticlines than on more open anticlines. Characteristically the "rolls" are asymmetrical, tending to overturn towards the crests of the anticlines. These features indicate drag folds produced by slipping of the beds during folding. This type of folding is much more pronounced in the relatively thin-bedded Mesozoic sandstone-shale-limestone series than in the massive Paleozoic limestone.

Where drag folding is most pronounced the folds have broken and the rocks slipped towards the crests of the anticlines along thrust faults that duplicate the beds (Pl. XXIX, A, B). Somewhat similar faults occur on the sharp turns around the margin of Tombstone basin, as the Lucky Cuss fault. Here, folding of the beds has been accompanied by faulting at a slight angle with the bedding which has in places duplicated strata.

STRUCTURAL CLASSIFICATION OF DEPOSITS

Factors other than structure influenced ore deposition, but here structure is considered. It is apparent from the preceding discussion that the different structures are interrelated. Many of the deposits are associated with more than one structure. Nearly all are associated with northeast-southwest fissures. The deposits may be classified, on the basis of the next most important associated structure, in the following groups:

Deposits associated with north-south (dike) fissures.

Deposits associated with faults.

Deposits associated with anticlines and "rolls."

Deposits with no obvious control other than northeast-southwest fissures.

It may be noted here that the most productive part of the Tombstone sedimentary section extends from somewhat below the top of the Naco limestone to a few hundred feet above the base of the Mesozoic beds. Much of the production has come from within 200 feet stratigraphically of the boundary of the Paleozoic and Mesozoic rocks. The productive Oregon-Prompter horizon is deeper in the Naco limestone. There are some deposits deep in the Paleozoic and some high in the Mesozoic section.

DEPOSITS ASSOCIATED WITH NORTH-SOUTH (DIKE) FISSURES

The deposits associated with north-south (dike) fissures have been most productive in the Empire-Contention-Grand Central dike zone, which crosses Tombstone basin and the anticlines, synclines, and "rolls" of the southeastern part of that basin. Others, notably the Tribute and the Tombstone Extension, belong to this type.

The north-south fissuring added to the fracturing produced by folding, and the rocks were still further disturbed by the introduction of the dikes. The final preparation of the ground was by the northeast fissuring.

The ore shoots are localized within the zone at the crossing of the northeast fissures. The ore is in fissures within the dikes, in the fissure zones occupied by the dikes, and in replaced beds, especially limestone beds extending away from the dikes. The ore shoots extend to several hundred feet stratigraphically above the Paleozoic-Mesozoic contact. There has been little prospecting of this type of deposit below that contact. The higher part of the Empire-Contention-Grand Central zone has been displaced down and to the east by the Tranquility fault and its branches.

The northeast fissures do not cross the north-south (dike) fissures directly on the strike but tend to swing into and follow them for some distance and to appear on the opposite side on the normal strike, but apparently offset. This change of attitude of the fissures in crossing the dikes is similar to the refraction of fissures in passing from one formation to another, described by Knopf[62] for the Mother Lode veins. It may also be due in part to actual offset by later movement in the zone.

DEPOSITS ASSOCIATED WITH FAULTS

The production from deposits associated with faults has come mainly from two groups of mines, one along the Lucky Cuss fault zone and the other along the Oregon-Prompter fault zone.

In both cases the deposits are associated with faults that have caused slipping along the beds near limestone-shale contacts. The Lucky Cuss zone is near the Paleozoic-Mesozoic contact, and the Oregon-Prompter zone is at the base of a shaly horizon deeper in the Naco limestone. The slipping resulted in fracturing the rocks and thickening or duplication of some of the beds, partly by faulting, partly by drag folding. Within the fault zones the ore shoots are at and near the crossing of northeast fissures. The fissures do not cross the fault zones on their normal strike but seem to swing into and follow the fault zone for some distance and appear on the opposite side with normal strike.

[62] Adolph Knopf, *The Mother Lode System of California* (U.S. Geol. Survey Prof. Paper 157, 1929), p. 24.

The deposits of the Oregon-Prompter zone are near, but only partly in, a reverse fault of large stratigraphic throw.

Deposits Associated with Rolls

"Rolls" or drag folds on the anticlines of the Tombstone basin are the sites of many ore shoots. Location of ore shoots on the "rolls" is influenced by several structures and combinations of structures so that prediction of shoots is not certain. The factors that localize ore shoots may be given in the following order, with due allowance for variation:

1. The overlying shale is a relatively impermeable cap.
2. The "novaculite" (siliceous shale) is the most brecciated and the most permeable rock.
3. The upper portion of the Naco limestone, the "novaculite" and the Blue limestone are chemically favorable to replacement.
4. Fissuring, brecciation, and permeability are generally greatest where the bends in the beds are sharpest. As the folds are not symmetrical, the sharpest bends may or may not be at the apexes. In some folds slipping along the beds produced permeable zones in a limb, down which mineralization extends for long distances.
5. After folding, the north-south (dike) fissuring and the intrusion of dikes further brecciated the rocks.
6. Northeast fissuring completed the preparation of the ground for mineralization. The mineralizing solutions rose through the northeast fissures and passed into the permeable zones.

The result of the several controls has been variable, but some generalizations may be ventured.

Mineralization is greatest in and near the northeast fissures and at or near the crests of anticlines (subject to exceptions). It is least at or near the troughs of synclines.

The most favorable beds are near the base of the "novaculite" and in the upper part of the Blue limestone. Other horizons in the "novaculite" as well as the limestone beds in the shale series above the Blue limestone are mineralized near strongly mineralizing fissures. Where the "novaculite" contains ore, the overlying Blue limestone may or may not be barren.

Mineralization extends outward along favorable beds from fissures crossing a roll. That from one fissure may meet that from another and thus form a continuous shoot along a roll. Such shoots have been followed for several hundred feet and are generally regarded as units, although really composed of several shoots.

Deposits Associated with Northeast Fissures Only

Deposits in northeast-southwest fissures have been productive in the quartz latite porphyry and in the Mesozoic sedimentary rocks. In the quartz latite porphyry they are simple fissure filling. In the Mesozoic sedimentary rocks mineralization is in the fissures and extends outward from the fissures along favorable beds.

In the Tombstone basin it is notable that the northeast-south-west fissures have been more productive in the anticlinal areas than in the synclinal areas (Pl. XXIX, A, B), but such are not the fissure deposits in their simplest form.

RELATION TO CAUSAL STRESSES

It is the purpose of this paper to present the relation of the ore shoots to structures without a discussion of the causal forces that produced either. It may be noted, however, that the structures probably resulted from stress from a north to northeasterly direction, and that the different structures can with some doubts and uncertainties be fitted to the idea of the strain ellipsoid.

The source of the metals is likely the same as that of the igneous rocks of the district, though the connection is inferred rather than obvious.

DISTRIBUTION OF METALS

Tombstone is essentially a precious-metal district. Of the production by value from 1879 to 1933, silver amounted to about 81 per cent and gold about 14 per cent of the total.[63] The remaining 5 per cent was mainly lead with some copper, manganese, and zinc.

The ores in different parts of the district range greatly in content of the different metals, and whether they were recovered or not depended somewhat on the degree of oxidation. Zinc has been recovered only from sulphide ores, and the distribution of the oxidized zinc minerals is little known. Manganese, on the other hand, has been recovered only from oxidized ores.

The distribution of metals suggests an area of most intense mineralization in the northeastern part of the district with a rough zoning outward. The most definite of the metal zones are the central gold zone and the marginal manganese-silver zone.

ACKNOWLEDGMENTS

During 1906 and 1911, F. L. Ransome carried on detailed mapping of the surface rocks of the district and much underground mapping for the U.S. Geological Survey. His report was not completed, and the work of the writers is a continuation of that study.

In 1936-37 James Gilluly, of the U.S. Geological Survey, contributed to the mapping of the Tombstone area and also to the interpretation of the geologic relations.

The writers are greatly indebted to the officials of the Tombstone Development Company, especially Ed Holderness; to R. T. Walker and E. P. Jeanes, of the United States Smelting, Refining, and Mining Company; and to C. M. d'Autremont and Harry Hasselgren, of the Tombstone Mining Company. J. H. Macia has

[63] M. J. Elsing and R. E. S. Heineman, *Arizona Metal Production* (Univ. of Ariz., Ariz. Bureau of Mines Bull. 140, 1936), p. 91.

been of great help to us and was similarly helpful to Dr. Ransome. During 1936-37 the Eagle Picher Lead Company made an examination of the area under the direction of George M. Fowler and generously contributed to the data collected. Acknowledgment is made of the contribution of Dr. C. A. Rasor, who assisted in the work.

BIBLIOGRAPHY

Blake, W. P., The Geology and Veins of Tombstone, Arizona, Am. Inst. Min. Eng., Trans., Vol. 10, pp. 334-45, 1882.
Church, J. A., The Tombstone, Arizona, Mining District, Am. Inst. Min. Eng., Trans., Vol. 33, pp. 3-37, 1903.
Ransome, F. L., Deposits of Manganese Ore in Arizona, U.S. Geol. Survey Bull. 710, pp. 96-103, 113-19, Pl. V, 1920.
Butler, B. S., Wilson, E. D., and Rasor, C. A., Geology and Ore Deposits of the Tombstone District, Arizona, Univ. of Ariz., Ariz. Bureau of Mines Bull. 143, 1938.

CERBAT MOUNTAINS[64]

BY ROBERT M. HERNON[65]

INTRODUCTION

Geography.—The Cerbat Mountains, in Mohave County, Arizona, extend for about 30 miles northward from Kingman, a town about 70 miles southeast of Boulder Dam. It is a desert range that attains altitudes of 5,000 to 7,000 feet and rises sharply for 1,500 to 3,500 feet above detritus-filled desert valleys. The erosion forms in this range are typical of granite and gneiss masses, except where remnants of lava flows cap mesas of the familiar southwestern type.

Water supply.—Water is not abundant in either the mountains or valleys. Some springs and wells are in volcanic rocks as at Kingman. The crystalline complex of the mountains has little primary porosity, and the small amounts of water generally found in it occur in fault fractures and joints. According to reports, wells in the detrital valley fills have yielded little water.

Literature.—The most extensive publication that deals with the Cerbat Mountains is by Schrader.[66]

Bastin[67] studied some of the rich silver ores during the secondary

---

[64] Paper prepared for, and originally presented at, the regional meeting of the A.I.M.&M.E. held at Tucson, Arizona, November 1-5, 1938.

[65] Assistant Professor of Geology, University of Arizona.

[66] F. C. Schrader, *Mineral Deposits of the Cerbat Range, Black Mountains, and Grand Wash Cliffs, Mohave County, Arizona* (U.S. Geol. Surv. Bull. 397, 1909).

[67] E. S. Bastin, *Origin of Certain Rich Silver Ores Near Chloride and Kingman, Arizona* (U.S. Geol. Surv. Bull. 750, 1924), pp. 17-39.

sulphide-enrichment investigations. Brief summaries of the geology and ore deposits have been given by others.[68]

Production.—The production[69] of the Cerbat Range through 1930 is given as follows:

Copper (lbs.)	Zinc (lbs.)	Lead (lbs.)	Gold	Silver	Total
2,900,000	95,587,344	55,350,000	$2,339,000	$5,038,000	$20,270,000

To this should be added approximately $170,000 for 1931-36, inclusive, and an unknown amount for some early production which, because of marketing conditions, was not credited to the Cerbat Range. The value of the total production is estimated at $21,000,000 to $25,000,000. Nolan[70] records that the mines of the Wallapai district produced 548,035 tons of ore valued at $13,955,473 during 1902-32.

The largest past producers by far have been the Tennessee and Golconda mines. The important producers at present are the Tennessee-Schuylkill and the Arizona-Magma mines near Chloride, and Keystone, Inc., which operates in Mineral Park and in and near the "Top of Stockton Hill" area. Some custom milling ore was produced in 1937-38 by the Minnesota-Connor Mine. Numerous other mines are yielding shipping ore and custom mill ore to small operators and lessees.

The larger mills include those of the previously mentioned main active operations, besides the Oro Plata mill (now idle), and the General Ores Reduction custom mill.

History.—Most of the mines of the Cerbat Mountains were discovered between 1863 and 1900. The metals sought in the earlier days were gold, silver, and lead. Rich silver chloride, silver sulphide, and native gold ores were exploited first. With cheaper transportation, base-metal ores were mined for lead with low silver. Subsequent improvement in milling methods led to exploitation of complex lead-zinc ores. The later history of the area is essentially the history of the Golconda and Tennessee mines, as they were affected by metal prices and marketing conditions and by milling methods.

The area reached its peak production in the years 1915-17, when the annual yield averaged nearly $3,000,000. This peak coincided with high metal prices. After the World War, production was small until 1936 when the Tennessee-Schuylkill Corporation began operations.

---

[68] R. T. Mason, *Mining in Northwestern Arizona* (Min. and Sci. Press, 1917), pp. 627-28.
N. H. Darton, *A Résumé of Arizona Geology* (Univ. of Ariz., Ariz. Bur. of Mines Bull. 119, 1925), p. 180.
W. Lindgren, *Mineral Deposits* (4th ed., 1933), pp. 578-79.
E. T. McKnight, *Mesothermal Silver-Lead-Zinc Deposits* (Am. Inst. Min. Eng., Lindgren Volume, 1933).
T. B. Nolan and others, *Mineral Resources of the Region around Boulder Dam* (U.S. Geol. Surv. Bull. 871, 1936), pp. 18-19.

[69] Morris J. Elsing and Robert E. S. Heineman, *Arizona Metal Production* (Univ. of Ariz., Ariz. Bureau of Mines Bull. 140, 1936), pp. 73, 95.

[70] *Op. cit.*

The Tennessee Mine had a small production in the early nineties. Schrader[71] states that for some time during the period 1897-1903, it yielded thirty to fifty carloads of concentrates per month, besides high-grade ores. In 1910 a new shaft was sunk. The Needles Mining and Smelting Company operated the property until 1916 and produced a considerable tonnage of excellent grade ores. Restricted, intermittent operations characterize the years 1917-36.

The Golconda Mine, which was developed later than the Tennessee, made a small production early in this century. A good zinc ore shoot was reported developed about 1907. From 1908 to 1917 the Golconda was exploited to the 1,100-foot level and reached its maximum production in the years 1915-17, a period that coincided with the greatest production from the Tennessee Mine. Attempts have been made to reopen the Golconda since the World War, but it has produced only from shallow workings.

GEOLOGY AND ORE DEPOSITS

The facts concerning the geology were derived from published data, information supplied by geologists who have worked in the district, and the writer's observations made mainly in the east-central part of the mineralized area. No detailed geologic map of the range exists.

Rocks.—The rocks of the Cerbat Range consist of pre-Cambrian crystalline rocks, later crystalline rocks of unknown age, and volcanic rocks of probable Tertiary and Quaternary age. Some of the Paleozoic and Mesozoic sedimentary rocks of the Colorado Plateau probably extended over the Cerbat area but were removed by erosion before the Tertiary volcanic activity.

The crystalline rocks of the Cerbat Range form a complex predominantly of granite with diorite and gabbro, all generally somewhat gneissic and intruded by pegmatite, medium-grained granite, diabase, granite porphyry, and lamprophyric dikes. Small- to medium-sized blocks of very dark schist (amphibolite) are locally common. All these rocks show various degrees of schistosity and represent two or more eras.

The rocks as classified by Schrader[72] are here summarized.

Quaternary	olivine basalt flows and detritus
Tertiary	thick volcanics
(? Mineralization)	
Mesozoic(?)	granite porphyry, diabase, minette and vogesite dikes
Pre-Cambrian	coarse-grained, porphyritic, gneissoid granites, granite altered to schist, diorite, amphibolite, graphite schist, pegmatite

Granitic rocks greatly predominate in the range. The lamprophyric dike rocks are locally termed "diabase," and much of the granite porphyry of local usage is actually porphyritic granite.

[71] *Op. cit.,* p. 54.

[72] *Op. cit.,* pp. 27-42, 49-118.

The pre-Cambrian rocks are slightly to strongly schistose, and the schistosity generally strikes northeasterly but locally north or northwesterly. Pegmatite commonly occurs in tabular masses along the pre-existing schistosity. Large masses of pegmatite crop out north of Kingman; one such mass is being exploited for feldspar.

The Mesozoic (?) rocks are of undetermined age. According to Schrader[73] they cut pre-Cambrian rocks and are older than the Tertiary volcanic rocks. The ore deposits are associated with rocks of this group in space and time. The vein faults cut across the members of the pre-Cambrian group of rocks and probably across the diabase, but monzonitic dikes and highly altered dikes with quartz phenocrysts intrude along the faults and are cut by the mineralized fissures. The lamprophyric dikes also intrude the vein faults and are mineralized. The diabase has been seen to form one wall of veins and appeared to be older than the vein faults. It closely resembles the diabase sills in the Grand Canyon series and in the Apache group of southeastern Arizona. All the rocks grouped as pre-Cambrian and Mesozoic (?) may show north to northwest sheeting that appears to pass into true schistosity where most intense; this sheeting and schistosity appears to be related to the northwest system of faults.

The prominent felsitic Broncho dike of the Cerbat mining district is said to cut off and offset northwest-striking faults and their vein filling. This dike may be related to the Tertiary rhyolitic volcanic rocks. A small, similar dike crops out in lower Cerbat Wash.

The Tertiary volcanics are limited to remnants around the margins of the Cerbat Range and to the crest in the extreme north and south parts of the range. According to Darton[74] they are principally rhyolite flows, tuffs, and agglomerates. The absence of veins of the Cerbat type in the volcanics and the presence of felsitic dikes cutting across the veins are evidence that the mineralization is of pre-Tertiary age. The volcanics are absent over the main mineralized area, however, and the strike of the veins is the same as in Oatman district, where the veins are younger than volcanics probably contemporaneous with the rhyolitic series of the Cerbat Range.

A considerable thickness of detritus occupies the valleys. Some of the older detritus is covered by Quaternary olivine-basalt lava which laps over on older bedrock.

Structure.—Pre-Cambrian and later structures are not well known because of the small amount of detailed mapping in the range. The older rocks and structures are cut by faults of northwest strike. These faults are of two directions at any one place and appear to represent relief by shearing. Striations generally indicate that movement along steeply dipping faults had a larger horizontal than vertical component. Some minor faults of about

[73] *Op. cit.*, p. 30.
[74] *Op. cit.*

50 degrees dip are striated parallel with the dip. That the rocks now visible were faulted under deep-seated conditions is indicated by clay gouge and finely crushed rock along the tight fissures; no open breccia is present except postmineralization breccia in the quartz veins. Tear fractures in the wall of the faults, attitude of striations, tightness of the faults, and the development of two directions of breaks, indicate stresses were mostly compressive and that the yield was mainly by shearing.

At least four main periods of movement are discernible along the vein faults. The initial break and one period of reopening were premineral in age; one main reopening occurred during mineralization; and movement occurred after mineralization was complete.

The most prominent direction of jointing and sheeting is northwesterly. Postmineralization cross faults are known at several places in the range.

Veins.—The veins were formed by solutions rising along the system of northwest fault fissures. It is estimated that the mappable veins would aggregate a total length of 100 miles and perhaps twice that length within the main mineralized area. Much of the vein matter is barren or of very low grade but locally is ore; narrow, noncommercial stringers of the valuable minerals may persist for long distances along or in barren vein filling.

The veins consist mainly of fine-grained quartz with pyrite, galena, sphalerite, and other minerals. Lindgren[75] classes them as mesothermal, pyritic galena-quartz veins of the Freiberg type, although they contain some gold. The veins generally do not exceed 5 to 10 feet in width, although locally, some are as much as 25 or 30 feet wide. Their ore shoots as a rule are 0.5 to 4 feet in width, though lenses attain widths of 6 to 14 feet in the largest ore shoots. The veins locally show a rough banding.

Mineralogy.—The minerals ordinarily seen in hand specimens of sulphide ore are quartz of three ages, pyrite of two or more ages, galena, and sphalerite. Bastin[76] records the following minerals in the rich silver ores; those marked with an asterisk are rare under the conditions indicated.

Oxidation products: cerargyrite, native silver,* copper pitch ore,* malachite,* native copper.

Secondary sulphide enrichment products: argentite,* proustite (very rare),* covellite,* chalcocite.

Primary (hypogene) minerals: quartz (generally gray and finely crystalline), manganiferous siderite,* calcite (white), pyrite, arsenopyrite, sphalerite, galena, chalcopyrite, tennantite,* argentite, proustite, pearceite,* polybasite.

Bastin emphasizes the arsenical nature of the high-grade silver ores. He notes that proustite is abundant in such ores and tends to occur with tennantite.

[75] *Op. cit.*, p. 578.
[76] *Op. cit.*, p. 35.

Bastin[77] found native silver near fractures and vugs showing oxidation. He notes that native silver ores grade directly into rich sulphide ores below and appear to be somewhat below the silver chloride ores.

Character of ores.—The unoxidized ore shoots are generally complex assemblages of galena, sphalerite, and gangue minerals, which carry gold, silver, and a small amount of copper. Indium is reported in some ores of the range. A few unoxidized ores are essentially gold-silver ores with normally low percentages of base metal. The greatest production, however, has been of lead-zinc ores with some gold and silver. Production statistics seem to indicate that ores with high-grade zinc carry the most gold and ores with high-grade lead the most silver, but the association does not appear to be a close one. Gold occurs with a bronzy pyrite according to Garrett,[78] geologist at the Tennessee Mine. Some gold may be associated with arsenopyrite. Silver occurs as argentite and sulpho salts in unoxidized ores, as previously described.

Mineralization.—The sequence of mineralization has not been determined in detail. Main stages are as follows: Some vein faults or parts of vein faults were intruded by dikes before mineralization began. Reopening followed with introduction of quartz along most of the length of faults, whether or not intruded by dikes. This produced quartz veins or lodes in places as much as 30 feet wide. Some pyrite crystals appear to be associated with this early quartz. Mineralization had probably largely ceased when reopening of the fault fissures occurred. Solutions brought in the valuable constituents of the veins, and quartz was reworked with apparently little further addition of silica. A weak reopening followed, with introduction of a little quartz as veinlets that cut the sulphides. Later reopenings produced quartz breccias and more gouge, but mineralization seems to have completely ceased.

Vertical or concentric zoning does not appear to be striking. Zinc is said to increase with depth, but high-grade lead is found in considerable amount in depth. Copper is said to increase slightly with depth. A type of horizontal zoning is found in some ore shoots: more or less vertical or steeply raking sections of a single ore shoot are characterized by high gold, high lead, or high zinc. These horizontal variations appear to be due to changes in stability of minerals with time, as acted upon by successive intermineral reopenings. Either sphalerite or galena may occur without the other; sphalerite appears to be the earlier mineral.

Wall-rock alteration is not extensive except where strong silicification of sheeted zones occurred in the early quartz stage. The granitic rocks appear to have silicified more readily than basic dikes or schist masses. Microscopic study of the wall rocks might show more extensive alteration than is apparent to the eye.

Localization of ore shoots.—In the main mineralized sections of the Cerbat Range the 100 or more miles of vein outcrops are

[77] *Op. cit.*, p. 36.
[78] S. K. Garrett, personal communication, 1938.

composed mainly of barren or very low-grade material. According to available production records, only two mines, the Tennessee and the Golconda, exceeded a total production of $1,000,000. While a great many mines have made appreciable productions, the geological conditions favorable for ore bodies of the size of the Tennessee and Golconda are rare. These two ore shoots were explored for vertical distances of 1,600 and 1,400 feet, respectively.

Schrader[79] noted that some ore shoots coincide with intersections or forking of veins. Many vein intersections, however, do not show ore shoots.

Ore shoots appear to be localized where changes of strike or dip of the vein faults gave rise to open spaces due to the reopening movements that occurred just before and during mineralization. Open space filling seems to have been most important as far as valuable vein minerals are concerned. Areas of faults choked by either clay gouge or greatly crushed rock were too tight for big ore shoots. No striking control of ore shoots by wall rock is known. One small shoot was seen to pinch out where the vein passed from granitic rock to dense black schist.

Oxidation.—Weathering of the veins is incomplete where the filling is highly siliceous, except along open fractures or where the vein is brecciated. High-grade sphalerite ore shoots or heavy pyrite streaks were more or less completely oxidized and leached. Galena, however, is often seen on natural outcrops. Water level is ordinarily at depths of 25 to 250 feet, but oxidation does not tend to be prominent for more than 30 to 100 feet, except along open fissures. Ground water is rich in chlorine, according to Bastin,[80] who found 80 parts per million in a stream near the town of Chloride.

Secondary enrichment.—Bastin[81] does not believe that secondary sulphide enrichment of silver and copper is important in rich silver ores. His microscopic studies indicate argentite, occurring in funguslike patches, to be the main secondary silver mineral. He found pearceite and abundant proustite intimately associated with primary sulphides to be probably primary.

Several veins, however, may have undergone considerable secondary enrichment. An exploited vein in Mineral Park shows small base-metal shoots with good silver content that dropped out below the third level. The narrow Alpha vein in the Cerbat district has a strong gossan at the outcrop. Schrader[82] noted silver sulphide, pyrite, galena, zinc blende, and chalcopyrite in Alpha ore. Chalcocite can be seen in some specimens. Ores mined recently had high copper and silver content and appeared to be secondarily enriched.

Regardless of whether the veins have been enriched primarily or secondarily in silver, available evidence does not indicate that

[79] *Op. cit.*, p. 51.

[80] *Op. cit.*, p. 18.

[81] *Op. cit.*, pp. 36-37.

[82] *Op. cit.*, p. 103.

high-grade silver can be expected to extend downward more than a very few hundred feet.

Gold has been enriched residually by leaching of zinc and iron from heavy sulphide ore shoots carrying relatively low primary gold. A thin zone of very rich gold ore is reported near the bottom of the oxidized zone in several veins. This may be secondary gold. Nature of gangue, ground-water chloride ion, common presence of pyrite, and persistent though only locally abundant manganese oxides are all favorable for gold enrichment. Some gold enrichment has occurred, but how much residual and how much chemical is unknown. Such gold ore shoots have been small, but some were spectacular. Many sections of veins that are very low grade in the sulphide zone have yielded small bodies of gold ore of shipping grade from the oxidized zone.

Summary.—The Cerbat Range is an area of numerous veins with mostly small ore shoots. The excellent grade ores and fair-sized shoots of several mines indicate the area to be important and worthy of study. The great need of the present is for a good topographic map of adequate scale and for a sufficiently detailed geologic map to bring out essential features. Many problems of structure, petrology, ore occurrence, and mineralogy are unsolved. Microscopic study of ordinary sulphide ores is needed. The exact manner of occurrence of gold and silver in ores of ordinary grade should be determined.

Acknowledgments.—The writer is indebted to G. M. Fowler, of Joplin, Missouri, for direction and for the opportunity to study part of the Cerbat area. Many local people facilitated the field work and gave information.

TENNESSEE-SCHUYLKILL MINE[83]

By S. K. GARRETT[84]

LOCATION

The Tennessee-Schuylkill Mine is at the western foot of the Cerbat Range, about 1 mile east of Chloride, in the Wallapai mining district, Mohave County, Arizona.

ROCKS

The rocks of the Wallapai mining district can be grouped as diorite gneiss, granite, quartz monzonite porphyry, rhyolite, and diabase. The oldest rock, diorite gneiss, has been intruded by granite, and both the diorite gneiss and the granite have been intruded by quartz monzonite porphyry. The rhyolite and diabase

[83] Paper prepared for, and originally presented at, the regional meeting of the A.I.M.&M.E. held at Tucson, Arizona, November 1-5, 1938.
[84] Geologist, Tennessee-Schuylkill Mine.

occur as dikes, some of which are in the same fissures as veins. In one place a diabase dike has been intruded along an earlier rhyolite dike.

VEINS

The fissure veins near Chloride can be grouped according to strike. One set strikes nearly north and the other about N. 25 degrees W.; the dip ranges from 35 degrees E. at the western foot of the range to 85 degrees W. near the crest. The progressive steepening toward the crest of the range may indicate overthrusting stresses as the cause of the fissuring.

The Tennessee-Schuylkill fissure vein, which can be traced for nearly 2 miles, strikes N. 5 degrees W. and dips 85 degrees NE.

Strong gouge is present on both the hanging and footwalls of the vein. There was some movement on the fissure after the formation of the vein.

At abrupt changes in strike, there is some horse tailing of the fissure, but there are no cross fissures.

ORE DEPOSITS

The Tennessee-Schuylkill deposits occur as a vein filling a fissure in the complex of diorite gneiss, granite, and quartz monzonite porphyry. The ore is in shoots which, above the 900-foot level, rake northward and between the 900- and 1,400-foot levels are nearly vertical (Pl. XXX).

Most of the ore shoots range from 35 to 300 feet in length and average about 5 feet in width.

ORE CONTROLS

The different wall rocks have not influenced the deposits; the ore filling is as wide in diorite gneiss as in quartz monzonite porphyry. The only recognized control is that of strike and dip of the fissure.

The four ore shoots in the Tennesee-Schuylkill Mine (Pl. XXX) occur where the vein has changed to a more than average northwesterly strike. The ore filling is wider on steep dips than on flat dips.

The combination of strike and dip control the rake of the ore shoots. A change to a northwesterly strike on a flat dip gives a pronounced northward rake, and a change in strike on a steep dip gives a rake that varies from slightly southward to vertical.

ZONING

There is marked horizontal zoning of the ore minerals in two of the ore shoots above the 900-foot level. The north limits of these two shoots contain principally galena and gold-bearing pyrite with practically no sphalerite. As the south limits of the shoots are approached, the galena and gold-bearing pyrite decrease, and sphalerite increases until, at the southern limits of the shoots, sphalerite is the only ore mineral present (Pl. XXX).

Little is known of the zoning below the 900-foot level other than a general decrease of galena and increase in sphalerite and crystalline pyrite with increased depth. On the 1,600-foot level a small amount of development along one of the ore shoots shows no galena but considerable sphalerite and crystalline pyrite.

MINERALOGY

The hypogene ore minerals are galena, fine-grained gold-bearing pyrite, and sphalerite. The gangue minerals are milky quartz, fine-grained chalcedonic quartz, crystalline pyrite, and arsenopyrite.

Supergene ore minerals, found to a depth of about 80 feet are: plumbojarosite, anglesite, cerussite, bromyrite, cerargyrite, native gold, and, rarely, native silver. The supergene ores are of little importance.

The paragenesis, determined megascopically, is milky quartz, sphalerite, galena, pyrite, and fine-grained chalcedonic quartz.

The sphalerite occurs as older "black-jack," and younger "rosinjack." Some galena shows a flow structure suggesting movement of the walls of the fissure after deposition. Argentite may account for the small amount of silver that the ore contains.

The pyrite is of two varieties. One variety occurs as well-crystallized cubes and pyritehedrons with no gold; the other is somewhat massive and fine grained and contains 0.3 to 15.0 ounces of gold per ton in the pure specimens. The gold in the pyrite is so finely divided that colors cannot be panned from a high-grade pyrite concentrate.

The fine-grained chalcedonic quartz occurs as fracture fillings in the sulphide ore.

MONTANA MINE, RUBY[85]

BY GEORGE M. FOWLER[86]

INTRODUCTION

A brief description of the geology of a limited area around the Montana Mine is presented in this paper. During the past few years a much larger area was studied in an attempt to find new ore bodies that could be worked in conjunction with this operation. At a later date it is hoped to present the results of this investigation as well as to give further details about the Montana Mine (Pl. XXXII).

The Montana Mine is in the Oro Blanco mining district, Santa Cruz County, Arizona, 5 miles north of the Mexican boundary and about 30 miles west of Nogales, Arizona.

---

[85] Paper prepared for, and originally presented at, the regional meeting of the A.I.M.&M.E. held at Tucson, Arizona, November 1-5, 1938.

[86] Consulting geologist, Joplin, Missouri.

The apex of numerous veins, as shown in Plate XXXII, is a prominent quartz outcrop now known as the Montana vein. This outcrop was located as a mining claim in the early seventies and was probably known to the Spaniards and Mexicans at a much earlier date, since one of their old trails from Mexico to the north passes within a few miles of the mine. In 1891 a small stamp mill was built in California Gulch on the east end of the Montana vein, and milling operations continued until 1893. Before the successful introduction of the flotation process, the complex nature of the ore retarded operations.

In 1916 the Goldfield Consolidated Mines Exploration Company acquired the property, built a concentrating mill which included flotation, and opened the mine from the surface to about the 250-foot level. It continued operations until the World War.

In 1928 the Eagle-Picher Lead Company acquired the property and undertook an extensive exploration campaign through diamond drilling and mining. It delayed extensive production until 1934, having in the interim completed a 400-ton concentrating mill and a water system 16 miles in length from the Santa Cruz River. The property has been in continuous operation since late in 1934 by the Eagle-Picher Mining and Smelting Company, a subsidiary of the parent company.

The mine is operated through a vertical shaft about 700 feet in depth with intermediate working levels at 100, 200, 300, 400, 525, and 660 feet from the collar. Square setting and back filling are necessary. Two raises, the Orem and Jenkins, connect the lower levels with surface pits where waste rock is obtained. The natural water level is a few feet below the 525-foot level, and the volume is small.

GEOLOGY

The rocks in and around the mine consist largely of andesite, quartz monzonite, conglomerate, diorite of various types and several ages, and rhyolite and allied rocks and tuffs. Small patches of shale occur in a few scattered areas. All the formations earlier than the Blue Ribbon diorite, described later, have been cut by intrusive stocks and dikes.

The andesite and quartz monzonite are thought to be the oldest formations in the district, but their age relation is uncertain, as they are some distance from the mine workings and have been little studied. The andesite covers a large area in and east of California Gulch, a few hundred yards east of the Montana vein. Quartz monzonite was found in a diamond drill hole under the Montana Mine workings at a depth of about 1,000 to 2,044 feet, the bottom of the hole. The upper part of this formation is leached and the crystals and groundmass show marked fracturing. Surface exposures of it are absent in the vicinity of Ruby.

It was relatively easy to establish the age relationship of the other rocks, as many thousand feet of mine workings and many diamond drill holes thoroughly explore them. The rocks in sequence are:

1. Oro Blanco conglomerate—named from the mining district in which the property is located. It is tentatively regarded as of Mesozoic age.

2. Ruby diorite—named from the mine settlement.

3. Sidewinder diorite porphyry—named by the miners because numerous dikes of it trend along or through the veins, or both.

4. Blue Ribbon diorite—named from the characteristic bluish coloring of its outcrop.

5. Rhyolite and other volcanics which occur as flows and tuffs.

The Oro Blanco conglomerate, the host rock for the ores at the Montana Mine, covers a large area in the western part of Santa Cruz County and is probably part of a similar formation that is widespread in south-central Arizona, from west of the Baboquivari Mountains to the Santa Rita Mountains east of the Santa Cruz River.

The conglomerate is characterized by the angularity and the coarseness of its constituent fragments. It seems difficult to classify this rock, as it combines characteristics of a conglomerate and a breccia. The fragments range between an inch and 12 inches in diameter. Small gravel is present only in sufficient quantity to fill the interstices. The color is reddish and grayish, with some dark-hued, greenish fragments that give the mass a variegated appearance in some places.

The late Fred E. Gregory, former chemist and geologist at the Montana Mine, studied the glomeratic material and his notes show that the fragments are with one exception of igneous origin, derived in part from surface and partly from deeper-seated types. Quartzite represents the exception noted. The cementing material is mostly silica and exceptionally calcium carbonate.

The relative abrasion of the fragments indicates that the plutonic rocks have been transported a shorter distance and the effusive and metamorphic rocks a greater distance.

Ruby diorite, the footwall of the Montana vein in parts of the workings, covers a large area south of the Montana Mine. It is darker, finer grained, and denser than the younger diorite. Fragments of conglomerate are included in this diorite, and dikes of Sidewinder and Blue Ribbon formations cut it. A crosscut on the 500-foot level, southeast of the Montana shaft, penetrates the Ruby diorite and a large inclusion of the conglomerate.

The Sidewinder diorite porphyry, which is younger than the veins, is widely exposed on the surface and underground. It occurs as dikes parallel to the Montana vein and as large, compact masses on all sides of the mine. The general strike of the dikes is easterly and the dip about 40 degrees to the north, which is about the same dip and strike as the Montana vein. The dikes range in thickness from a few feet to more than 100 feet. Some dikes forced their way into fissures on one or both sides of the numerous splits of the vein system and between segments of the veins, which further complicated the structural pattern of these ore deposits.

On the 100-foot level south of the main shaft a little ore was mined from a fissure in crests of a Sidewinder dike. The material included angular and rounded pieces of ore and barren fragments of conglomerate and diorite porphyry that had been pushed ahead of the dike. In another place a large segment of the Rough and Ready vein was pushed aside laterally by a dike. The segment then broke into several large pieces and Sidewinder dike material filled the intervening spaces. This ore body, more than 100 feet in height, is being mined at the present time (Pl. XXXIII).

The Blue Ribbon diorite is the youngest igneous formation in the mine workings. It is well exposed in front of the guest house at the mine. It is fine grained and weathers rapidly to a soft bluish color. Dikes are present in the mine workings and over a large area south of the mine. The dikes in the mine are designated on the map as the Montana, Rough and Ready, and Philadelphia. The first two, about 30 feet wide, strike nearly north and cut the veins at right angles.

The Oro Blanco conglomerate rests on a weathered quartz monzonite surface under the mine workings, a relationship which may be widespread.

STRUCTURAL DEFORMATION AND ORE DEPOSITS

Structural deformation and the ore deposits are so closely related that it seems best to discuss them together. It is apparent from the surface geology that deformation was regional in character and of greatest intensity in the vicinity of Ruby. Oro Blanco conglomerate occupied depressions in the region to a depth of many hundred feet. The conglomerate and the younger formations were intruded successively by the Ruby diorite, Sidewinder diorite porphyry, and Blue Ribbon diorite.

The conglomerate-Ruby diorite contact presents a jagged surface, as numerous projections of the diorite extend a few feet into the conglomerate. In the mine a shear zone, known as the Montana vein, trends east-west along this contact. Its width ranges from mere stringers to more than 40 feet. All the ore bodies are in shear zones along this contact or in the conglomerate. Some shear zones branch with barren country rock between branches. Wide ore bodies in the conglomerate change to small stringers in the diorite. It seems probable that the diorite was resistant to deformation, whereas the conglomerate shattered intensely and made ideal ore reservoirs.

The shear zones dip from 40 degrees N. to almost vertical. In plan they are arranged in echelon with offsets ahead and to the right. The most important veins are the Montana and the Rough and Ready (Pl. XXXII).

The Montana vein can be traced on the surface for about 3,000 feet, in part as bold outcrops of quartz several hundred feet long and 50 feet wide, as exposed a few hundred feet south of the Montana shaft and around the Jenkins shaft (Pl. XXXII).

The outcrops of the Rough and Ready vein are numerous small, barren fissures, some of which contain quartz stringers. The vein was practically barren to a depth of about 300 feet.

The ore shoots in the Montana and Rough and Ready veins pitch about 45 degrees westward. (See Pl. XXXV which shows a longitudinal section of the Montana vein.) It seems probable that the center segment of the Montana vein dropped on the Montana dike, as the stopes on opposite sides of the dike have a vertical difference of more than 100 feet.

On the surface, in the mine workings, and in drill holes are many minor veins contiguous to the Montana-Rough and Ready vein system. Some of these on Company-owned property have been prospected with negative results.

The earliest mineralization at the Montana Mine was barren, milky-colored quartz which was deposited in the shear zones and replaced some of the country rock. It ranges in thickness from minute bands to veins many feet wide. Subsequent deformation shattered the brittle quartz and country rock and made excellent reservoirs for the ore-bearing solutions that replaced most of the quartz and country rock to form some of the ore bodies. Zones of intense deformation and little movement, such as the loci of torsional stresses, became particularly favorable ore reservoirs. Some of the zones show a fracture pattern with the major ore stringers trending with the vein and minor stringers crossing it at oblique angles.

Postore faulting had only a minor influence in the ore bodies. Large faults follow the footwalls, and in some cases the hanging walls, of the vein system. A few faults parallel the veins with little displacement. Dikes of Blue Ribbon age were displaced a few feet (Pl. XXXII).

The primary ore minerals at the Montana Mine are largely blende and galena. Chalcopyrite, pyrite, and tetrahedrite are of common occurrence everywhere in the mine. The tetrahedrite carries gold and silver. Alteration products of these minerals occur very sparingly in the upper levels.

PRODUCTION

The earlier operators of the Montana Mine left only meager production records. The Arizona Bureau of Mines records show the Goldfield Consolidated Mines Exploration Company production for 1917 and 1918 as:

Lead	1,250,000 pounds
Zinc	1,300,000 pounds
Gold	$20,000
Silver	$43,000
Total values	$265,000

The records since the present operators took charge of the property are shown in Table 5.

TABLE 5.

	Tons milled	Ounces		Per cent		
		Gold	Silver	Lead	Zinc	Iron
1928-30, inclusive	79,686	......	......	......	......	......
1934	36,963	.075	6.14	5.22	5.71	1.85
1935	129,374	.07	5.77	4.43	4.41	2.01
1936	141,768	.06	5.21	3.51	3.50	1.80
1937	143,004	.053	5.06	2.94	2.91	1.97
1938 (to July 1)*	70,410	.054	4.68	3.26	3.62	1.89
	521,519†	......	......	......	......	......

* Assays shown for 1938 production are subject to correction.
† The copper content in this ore was approximately 0.30 per cent.

ACKNOWLEDGMENTS

A number of persons have worked on the progressive study of these ore deposits, particularly J. P. Lyden, J. M. Conrow, Neil O'Donnell, and Paul R. Murphy, who made many of the geologic observations. Grover J. Duff, E. D. Morton, and their assistants at the mine lived with the problem and offered valuable suggestions that are recorded in this paper. The Arizona Bureau of Mines made several rock determinations and co-operated in many ways.

MAMMOTH MINING CAMP AREA
PINAL COUNTY, ARIZONA[87]

By Nels Paul Peterson

INTRODUCTION

The Mammoth mining camp area, which forms part of the Old Hat mining district in Pinal County, Arizona, is about 50 miles northeast of Tucson. The productive claims in the area were located by Frank Schultz between 1879 and 1882. Production began in 1886 and to the end of 1936 amounted to $5,204,000, most of which was gold. In 1916 and 1917 most of the molybdenum produced in the United States came from the Mammoth mines.

ROCKS (PL. XXXVI)

The oldest formation in the area is the coarse-grained, porphyritic Oracle granite. It is probably the same granite that is overlain by the pre-Cambrian Apache group about 10 miles north of Mammoth. Dikes and irregular bodies of aplite and andesite porphyry intrude the granite.

[87] Paper prepared for the regional meeting of the A.I.M.&M.E. held at Tucson, Arizona, November 1-5, 1938. For a more detailed description of this area, see *Arizona Bureau of Mines Bull. No. 144* (Univ. of Ariz., 1938).

The northern part of the area is covered by a series of lava flows consisting chiefly of basalt flow breccia with some latite and tuff. The lava flows are tilted 45 to 65 degrees NE. Both the granite and lavas are intruded by dikes and sills of rhyolite and a breccia composed of fragmental material derived from all the earlier rocks in a matrix of rhyolite. The largest intrusive body is a sill of rhyolite and breccia intruded between the granite and the overlying lava flows in the southern part of the area. The Mohawk-New Year Mine is entirely within this sill. Gila conglomerate and recent alluvium cover much of the mapped area; the conglomerate is older than the last period of deformation that affected the region.

Structure (Pls. XXXVII and XXXVIII)

Most of the fractures are premineral. Of these early fractures only the mineralized faults effected much displacement. The most important postmineral faults are the Turtle and Mammoth faults. The Turtle fault strikes N. 70 degrees E. and forms the boundary between the granite and the lava flows, which have been depressed relative to the granite. The Mammoth fault strikes N. 22 degrees W. and depressed the eastern half of the area. It is younger than the Gila conglomerate.

Ore Deposits

The veins occupy fault fissures of general northwestward strike and steep dip. Movement along them continued at intervals until after the supergene alteration of the ores. Except in the Collins Mine where sulphides occur on the lower levels, supergene oxidation extends below the permanent water table, which is the lower limit of mining development. The scarcity of pyrite, however, largely prevented transportation and enrichment of the metals.

The mineralization is divisible into five stages (Fig. 11). The first three stages, separated only by renewed movement on the vein fault, consist of quartz, specularite, chlorite, pyrite, sphalerite, galena, fluorite, gold, and chalcopyrite. The fourth stage began with leaching along certain channels, followed by deposition of wulfenite and vanadium minerals. The fifth-stage minerals include carbonates, silicates, and sulphides of lead, zinc, and copper, all of which are later than the molybdenum and vanadium minerals. The minerals of the fourth and fifth stages are of a type generally considered as supergene. Those of the fifth stage have clearly resulted from the supergene oxidation of the early hypogene minerals. The origin of the molybdenum and vanadium minerals is not so clear, since no primary source of these metals has been found. The most logical explanation for the mineral relationships in the later stages is that the molybdenum and vanadium minerals are of hypogene origin but were deposited by a different type of solution from those that deposited the earlier hypogene minerals.

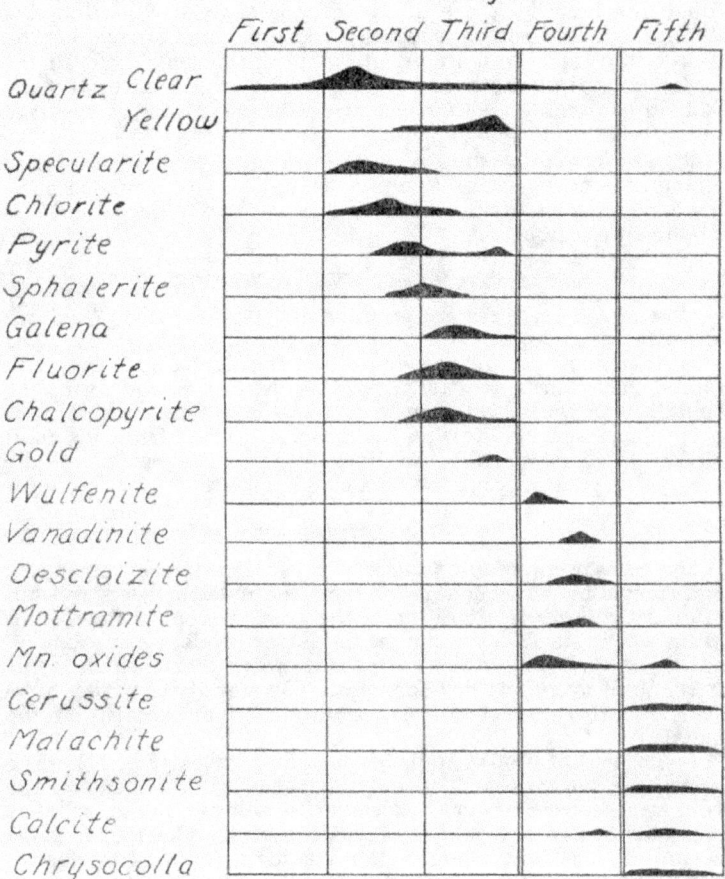

Figure 11.—Paragenetic relation of the vein minerals, Mammoth Mine, Pinal County.

The assemblage of minerals and the general character of the veins are both indicative of high temperature deposition at shallow depth.

BIBLIOGRAPHY

Rickard, T. A., Vein Walls, Amer. Inst. Min. Eng., Trans., Vol. 26, pp. 214, 233, 234, 1896.

Blake, Wm. P., Report of the Governor of Arizona, pp. 188-90, 1901.

Stewart, S. O., The Old Hat Mining District, Pinal County, Arizona, Mining and Engineering World, Vol. 36, p. 952, 1912.

Hess, Frank L., Molybdenum, U.S. Geol. Survey Bull. 761, p. 15, 1924.

Wilson, Eldred D., Arizona Lode Gold Mines, Univ. of Ariz., Ariz. Bureau of Mines Bull. 137, pp. 170-74, 1934.

Peterson, Nels Paul, Geology and Ore Deposits of the Mammoth Mining Camp Area, Arizona, Univ. of Ariz., Ariz. Bureau of Mines Bull. 144, 1938.

CHILDS-ALDWINKLE MINE, COPPER CREEK, ARIZONA[88]

BY TRUMAN H. KUHN[89]

LOCATION

The Childs-Aldwinkle Mine is at Copper Creek, in the western part of the Galiuro Mountains near the center of the Bunker Hill mining district. The camp is accessible by 10 miles of road that branches eastward from the Tucson-Winkelman highway at Mammoth.

HISTORY AND PRODUCTION

The Bunker Hill district was discovered in 1883 by prospectors who were searching for gold and silver. Mining was attempted in 1897-98 by the Table Mountain Copper Company; in 1903-17 by the Copper Creek Mining Company and its successors, the Minnesota-Arizona Copper Company, and the Copper State Mining Company; and in 1907-9 by the Calumet and Arizona Mining Company.[90] Between 1917 and 1933 a little work was carried on by lessees. Production from 1905-16 was approximately 700,000 pounds of copper and $35,000 worth of silver, with a total value of $150,000.[91]

In 1933 the Arizona Molybdenum Corporation, headed by W. C. Rigg, purchased the Childs-Aldwinkle property and began mining molybdenum and copper ore. According to Mr. Rigg, the production of this mine from 1933 to July 1, 1938, was 6,454,321 pounds of molybdenum sulphide, 5,519,140 pounds of copper, 473 ounces of gold, and 19,167 ounces of silver.

ROCKS

The canyon of Copper Creek, where it crosses the district, forms a small basin floored mainly by light gray granodiorite which in-

---

[88] Paper prepared for, and originally presented at, the regional meeting of the A.I.M.&M.E. held at Tucson, Arizona, November 1-5, 1938.

[89] Graduate student, University of Arizona.

[90] Historical data largely from unpublished notes of J. B. Tenney.

[91] M. J. Elsing and R. E. S. Heineman, *Arizona Metal Production* (Univ. of Ariz., Ariz. Bureau of Mines Bull. 140, 1936), p. 99.

truded limestone of undetermined age, probably in Laramide time. It is partly covered by two lava flows, the younger of which caps Sombrero Butte, Table Mountain, and other prominent elevations.

STRUCTURE

The main structural features in the vicinity of the mine are strong east-west vertical faults and fractures; indistinct north-south vertical faults; a northwestward-trending fault with steep southwestward dip; and pipelike masses of breccia.

Two of the east-west faults are traceable, with a fair degree of certainty, through the mine (Pl. XXXIX). One separates the main ore body from its southern branch, and the other is associated with the northern branch. Both may be traced on the surface as part of a large fault zone.

On most of the levels an indistinct north-south vertical fault is apparently associated with the north-south elongation of the ore body. It has little gouge and is difficult to trace through the breccia. Other fractures and faults generally displace it only to a small extent.

A fault on the 700- and 800-foot levels strikes N. 60 degrees W., dips 65 degrees SW., and passes through the main and the northern branch ore bodies. It has not been recognized on the surface.

The pipelike mass of breccia consists of blocks of altered granodiorite, cemented with gouge, gangue, and ore minerals. The blocks range from an inch to 15 feet in diameter and average 6 to 12 inches. They are relatively small and closely spaced in the upper portions of the pipe.

The breccia pipe is of oval plan and almost vertical. It is bounded in most places by an irregular but rather sharp contact with massive granodiorite.

ORE DEPOSIT

The Childs-Aldwinkle deposit, like most others in this district, is in a breccia pipe. The entire breccia mass, though it may be mineralized, is not all ore. The ore breccia in the central portion of the pipe grades outward into breccia that contains only pyrite in the lower part of the mine. In the upper 200 feet of the pipe the entire breccia was ore, but the outer rim was considerably richer than the core.

The ore minerals are confined to the cementing material where they occur as disseminations and lining of fractures. Some branches or offshoots from the main ore body have been followed, but others are too small to mine away from the main workings. There are two distinct oval-shaped outcrops of ore at the surface (Pls. XXXIX and XL) of which the northern was 270 by 150 feet and the southern 220 by 100 feet in diameter. They diminished about equally in relative size to a depth of 340 feet. For the next 50 feet, the northern ore body continued to diminish gradually, whereas the southern body decreased to one sixth the size of the northern or main ore body. From this point downward

Plate XL.—Looking west at glory holes of Childs-Aldwinkle Mine.

the exact location of the southern ore body is unknown, but small masses of ore on the main levels suggest that it continues downward and joins the main ore body about 700 feet below the surface. At a depth of 800 feet the northern body pinched out. Below this point the breccia contains molybdenite, but as small, scattered grains.

At a depth of 350 feet, the main ore body branches as indicated in Plate XXXIX. In places it is connected with the smaller or southern branch by low-grade ore. At a depth of 760 feet, the northern branch increases in size and changes from vertical to a dip of 65 degrees E. At the present bottom of the mine, 850 feet below the surface, the ore follows a north-south fault along which the oval outline becomes much elongated. On the 700-foot level a mineralized zone 6 to 12 inches wide, high in molybdenite and bornite, is associated with the north-south fault.

Molybdenite, bornite, and chalcopyrite are the principal ore minerals. Treatment of the copper concentrates yields small amounts of gold and silver as well as arsenic, antimony, and traces of bismuth. Near the surface are oxidized minerals, molybdite, malachite, azurite, cuprite, and small amounts of native copper. Copper is very subordinate to molybdenum in the upper levels. Below about 650 feet bornite and chalcopyrite are important except in the lowest levels where the amount of copper diminishes considerably. The average molybdenite content changes very little through the mine, and the amount of pyrite remains fairly constant.

The nonmetallic gangue minerals show the greatest variation. Near the surface quartz and calcite are the principal gangue minerals. On the main haulage level (300 feet below the surface) crystalline quartz is still abundant, but there are also crystals of orthoclase, with quartz in the open cavities, and calcite and gypsum. At a depth of about 600 feet a large amount of orthoclase and biotite, accompanied by numerous crystals of apatite, is present. Quartz is still present but in lesser amounts. On the 800 level orthoclase has diminished with a corresponding increase in biotite. At the present bottom of the mine (850 feet) the gangue is mainly coarsely crystalline quartz and biotite, with little orthoclase.

With increased depth the molybdenite crystals increase in size, changing from 0.04 to 0.12 inch in diameter to rosettes 1.8 inches in maximum diameter.

WALL-ROCK ALTERATION

Although the country rock surrounding the breccia pipe appears fairly fresh, the ferromagnesian minerals have in part gone to chlorite and the feldspars to sericite. The alteration of the iron-bearing minerals to hydrous ferric oxide gives the outcrops their characteristic reddish color. The ore breccia in many places is cemented almost entirely with alteration minerals, particularly on the 600-foot level where the breccia blocks are held together with a green, finely divided chlorite. In the upper levels alteration is less evident, but sericite is present. Chlorite and sericite occur in the lower portion of the mine in approximately equal amounts.

OCTAVE MINE[92]

By ELDRED D. WILSON

SITUATION

The Octave Mine is at Octave in the Weaver district of southern Yavapai County. It is reached by about 10 miles of road that leads eastward from Congress Junction, a station on U.S. Highway 89 and the Phoenix branch of the Santa Fe Railway.

HISTORY AND PRODUCTION

This deposit probably became known in the sixties, shortly after the discovery of the Rich Hill placers, but, as a large

[92] Acknowledgments are due B. R. Hatcher, Carl G. Barth, Jr., M. E. Pratt, and A. E. Ring for much information upon the Octave area. Paper prepared for the regional meeting of the A.I.M.&M.E. held at Tucson, Arizona, November 1-5, 1938.

part of the gold was not free milling, little work was done upon
it until the advent of the cyanide process. During the late nineties, according to local reports, a group of eight men purchased
the property and organized the Octave Gold Mining Company.
Between 1900 and 1905 the vein was mined to a depth of about
2,000 feet on the incline and for a maximum length of 2,000 feet
along the strike. This ore was treated in a forty-stamp mill
equipped for amalgamation, table concentration, and cyanidation.
The total gold and silver production during this period is reported
to have been worth nearly $1,900,000. In 1907 the property was
bought by a stock company that built an electric power plant at
Wickenburg, 11 miles away, and electrified the mine and mill.
This company, however, failed to work the mine at a profit and
ceased operations in 1912. In 1918 a group of the stockholders
organized the Octave Mines Company, with H. C. Gibbs, of
Boston, as president and carried on development of the Joker
workings until 1922. In 1928 the Arizona Eastern Gold Mining
Company, Inc., was organized. This company operated a 50-ton
flotation plant on ore already blocked out in the Joker workings.
After obtaining approximately $90,000 worth of concentrates
from 9,100 tons of ore, operations were suspended in 1930. This
ore contained nearly equal proportions by weight of gold and
silver.[93]

The production from 1895 to 1925, inclusive, has been reported as
$1,825,000 worth of gold and $75,000 worth of silver, a total value of
$1,900,000.[94]

In 1934 the American Smelting and Refining Company obtained
control of the property, equipped the mill late in the year, and
began production from the Joker workings. According to figures
published by the U.S. Bureau of Mines, the yield during 1934-36
was as follows:

1934	2,636 tons
1935	23,951 tons
1936	22,107 tons (estimated)

The ore mined during 1936 was reported to average 0.364 ounce
of gold and 0.464 ounce of silver per ton, 0.240 per cent lead, and
0.030 per cent copper.

TOPOGRAPHY AND GEOLOGY

Here the rugged southwestern front of the Weaver Mountains
rises abruptly for more than 2,000 feet above the desert plains.

The Octave Mine is at an altitude of 3,300 feet on a narrow
pediment at the southwestern base of the range and 2 miles
south of Rich Hill. This area is dissected by several southward-
trending arroyos of which Weaver Creek, ¼ mile west of Octave,
is the largest. Water for all purposes is brought through 7 miles
of pipe line from Antelope Spring.

---

[93] Oral communication from H. C. Gibbs and M. E. Pratt.

[94] M. J. Elsing and R. E. S. Heineman, *Arizona Metal Production* (Univ.
of Ariz., Ariz. Bureau of Mines Bull. 140, 1936), p. 103.

The pediment and mountains immediately northeast of Octave consist mainly of grayish quartz diorite with included lenses
of schist and northeastward-trending dikes of pegmatite, aplite,
and altered, fine-grained basic rock. Examined microscopically
in thin section, the basic rock is seen to be an altered diabase.
The pegmatite and aplite dikes are earlier than the veins, and the
diabase dikes cut the veins. As shown by Mr. Barth's geologic
and relief model of the district, the dike systems are rather complex.

Mineral Deposit

The main Octave vein occurs within a fault fissure that strikes
N. 70 degrees E. and dips from 20 to 30 degrees NW. Cleavage
in its gouge indicates that the fault was of reverse character.
This fissure, which is traceable on the surface for several thousand
feet, has been intersected by three or four systems of postvein
faults of generally less than 100 feet displacement.

As seen in the Joker workings, the vein filling consists of
coarse-textured, massive to laminated, grayish white quartz together with irregular masses, disseminations, and bands of fine-
grained pyrite, galena, and sparse chalcopyrite. Most of the gold
is contained within the sulphides, particularly the galena, and
comparatively little is reported to occur free. According to Mr.
Pratt, the pure galena generally assays more than 100 ounces of
gold per ton, the chalcopyrite 8 to 25 ounces, and the pyrite from
3½ to 7 ounces. A band, from a few inches to a foot thick, of
barren, glassy, bluish gray quartz commonly follows one of the
vein walls.

The width of the vein ranges from a few inches to 5 feet and
averages approximately 2¼ feet. In places it narrows abruptly
and forms branching stringers. Locally, a parallel vein occurs
about 25 feet away from the main vein, but its relations have not
been determined.

Wall-rock alteration along the Octave vein consists of sericitization, silicification, and carbonatization. The deposit is of mesothermal type and presumably of Laramide age.

The Joker workings include a shaft 1,250 feet deep on the incline and several thousand feet of drifts. Most of the large stopes
are below the 300-foot level and within a distance of 800 feet east
of the shaft. A few smaller stopes are west of the shaft.

The old Octave workings, which extend to a depth of approximately 2,000 feet on the incline, were being unwatered in 1938.
They include four shafts with many thousand feet of drifts and
extensive stopes on three ore shoots. A sketch section of these
workings has been published by Nevious.[95]

---

[95] J. N. Nevius, "Resuscitation of the Octave Gold Mine." *Min. and Sci.
Press*, CXXIII (1921), 122-24.

ARTILLERY MOUNTAIN MANGANESE DISTRICT, MOHAVE COUNTY,
ARIZONA[96]

BY

SAMUEL G. LASKY[97] AND BENJ. N. WEBBER[98]

GENERAL FEATURES

During the past ten years, investigations of manganiferous
sediments on the west side of the Bill Williams River near Alamo
crossing have proved the presence of large reserves of low-grade
material. Alamo is 46 miles by road west of Congress Junction and
about equally distant northwest of Aguila.

The manganese-bearing formation underlies an area of possibly
25 square miles in the dissected valley between the Artillery and
Rawhide mountains. The richest and most extensive croppings
lie along the northeast side of this area among the hills that form
the southwest flank of the Artillery Mountains (Fig. 12). Prominent exposures in Secs. 32 and 33, T. 12 N., R. 13 W. were explored during the World War, and several cars of ore from Sec. 33
are said to have been shipped to Alabama in 1928. In 1929 the
Chapin Exploration Company of Duluth, Minnesota, and in 1930
the Arizona Manganese Corporation acquired large holdings and
explored parts of the area. At present, all the principal holdings
except those in the Price block, owned by the Arizona Manganese
Corporation, are controlled by the M. A. Hanna Company of
Cleveland, Ohio, under long-term leases, and an extensive dia-
mond-drilling program is under way.

Short descriptions of the deposits have appeared from time to
time,[99] and in the spring of this year (1938) S. G. Lasky was
assigned to make a more detailed examination of the occurrence.
Three months were spent in the field, and a detailed report is in
preparation. B. N. Webber, who has been intimately associated
with the development of the area since 1929, co-operated during
the survey, particularly with respect to the area covered by the
holdings of the M. A. Hanna Company.

GEOLOGY

The manganese-bearing formation, tentatively considered to be
of early Pliocene age, crops out roughly in the shape of an elon-

---

[96] Published by permission of the Director of the Federal Geological Survey. Paper prepared for the regional meeting of the A.I.M.&M.E. held
at Tucson, Arizona, November 1-5, 1938.

[97] Associate geologist, U.S. Geological Survey, and member A.I.M.E.

[98] Field geologist, M. A. Hanna Company, and member A.I.M.E.

[99] E. L. Jones and F. L. Ransome, *Deposits of Manganese in Arizona* (U.S.
Geol. Survey Bull. 710 D, 1920). E. D. Wilson and G. M. Butler, *Manganese Ore Deposits in Arizona* (Ariz. Bur. Mines Bull. 127, 1930), pp.
71-76. D. W. Woodbridge, "An Addition to Domestic Manganese Reserves," *Eng. & Min. Jour.*, CXXXV (1934), pp. 459-60. D. F. Hewett
et al., *Mineral Resources of the Region around Boulder Dam* (U.S. Geol.
Survey Bull. 871, 1936), pp. 81-83.

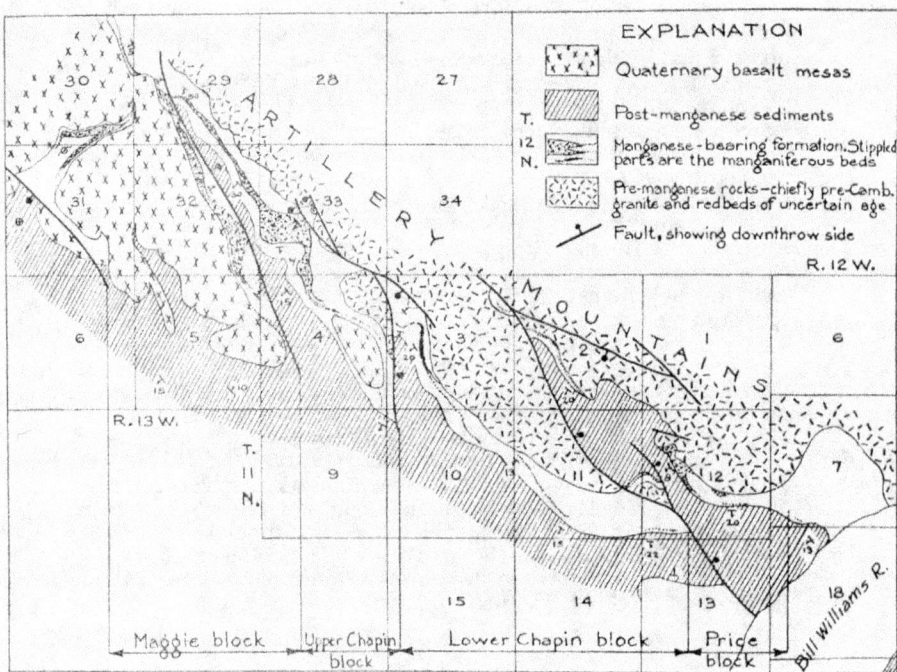

Figure 12.—Generalized geologic map of the main part of the Artillery Mountain manganese district, Arizona.

gated U, about 7 miles long and 3½ miles wide, along the flanks and axis of a syncline that extends northwest. The base of the U lies along the Bill Williams River. The arms of the U, along the flanks of the fold, lap against the older rocks of the Artillery and Rawhide mountains and here and there pinch out between these older rocks and some unconformable younger formations. The older rocks in the Artillery Mountains include chiefly pre-Cambrian granite, gneiss, and schist; red beds of uncertain age that locally rest on the pre-Cambrian rocks elsewhere thrust over them; and Tertiary volcanics. In the Rawhide Mountains the older rocks include chiefly the pre-Cambrian and rocks of the red-bed formation. The rocks younger than the manganese-bearing formation include mainly the Temple Bar (?) conglomerate, which occupies the central part of the syncline and rests unconformably upon the manganese-bearing formation, and Quaternary basalt flows that overlie the several other formations in parts of the Artillery Mountains.

The manganese-bearing formation was deposited in a down-faulted basin between the Artillery and Rawhide mountains and consists of alluvial fan and playa deposits derived from the rocks of these mountains. It is composed largely of compacted sands, silt, and clay, in part gray but generally pink to red, with some extensive conglomerate zones. The clays locally are gypsiferous, and layers of limestone crop out at various horizons at widely separated places. The lithology of the formation in different parts of the basin reflects the lithology of the older rocks that constituted the immediate source of the sediments, and in the area where the principal manganiferous beds crop out the formation contains much arkosic and volcanic material derived from the pre-Cambrian granite and the Tertiary volcanics. Some tuff layers, however, may have been deposited directly from volcanic sources. The thickness of the formation is highly variable because of the unconformable surfaces that limit it at top and bottom; the maximum thickness measured is 575 feet, but the actual maximum may be much greater. Renewed faulting along the old basin-and-range faults, particularly along the Artillery Mountains, has broken the original continuity of the formation (Fig. 12).

MANGANESE DEPOSITS

The manganiferous beds within the manganese-bearing formation differ from the other beds of the formation primarily in having a pigment and cement of black manganese oxides in place of and in addition to the red iron oxides. They are present at various horizons within the formation and include all the types of sediments of which it is composed. They constitute continuous manganiferous zones as much as 85 to 90 feet thick stratigraphically, and elsewhere they form zones as much as 350 feet thick composed of manganiferous beds separated by nonmanganiferous ones. Individual zones are traceable along the strike continuously for as much as 3 miles, and there is good probability that the various exposures shown in Figure 12 may be continuous along some devious path below the surface for more than twice that length.

Other manganiferous deposits in the area include similar stratified material in the old red beds of the basement rocks; faults and fissure zones in the basement rocks cemented with manganese oxides and older than the main manganese-bearing formation; and similar faults and fissures that cut the Temple Bar (?) conglomerate and the Quaternary basalts; but the available manganese in these deposits seems small.

It can be confidently stated, without at this time detailing the overwhelming evidence for this conclusion, that the manganese was originally deposited with the sediments, although the manganese minerals have since been reworked by circulating waters. The ultimate origin of the manganese is not known, however; field evidence clearly proves that some was derived from the deposits in the old red beds, certainly in part as detrital material, but the facts are as yet insufficient to evaluate this source, nor is the source of the manganese in the old red beds known.

Reserves

From numerous outcrops manganiferous beds are known to be continuous over long distances, and a number of drill holes have partially explored these beds beyond the outcrops, but the deposits have not been sufficiently prospected to make accurate estimates of the tonnage or grade possible. It is known, however, that the district contains many millions of tons of ore that run from several per cent to 20 per cent manganese.

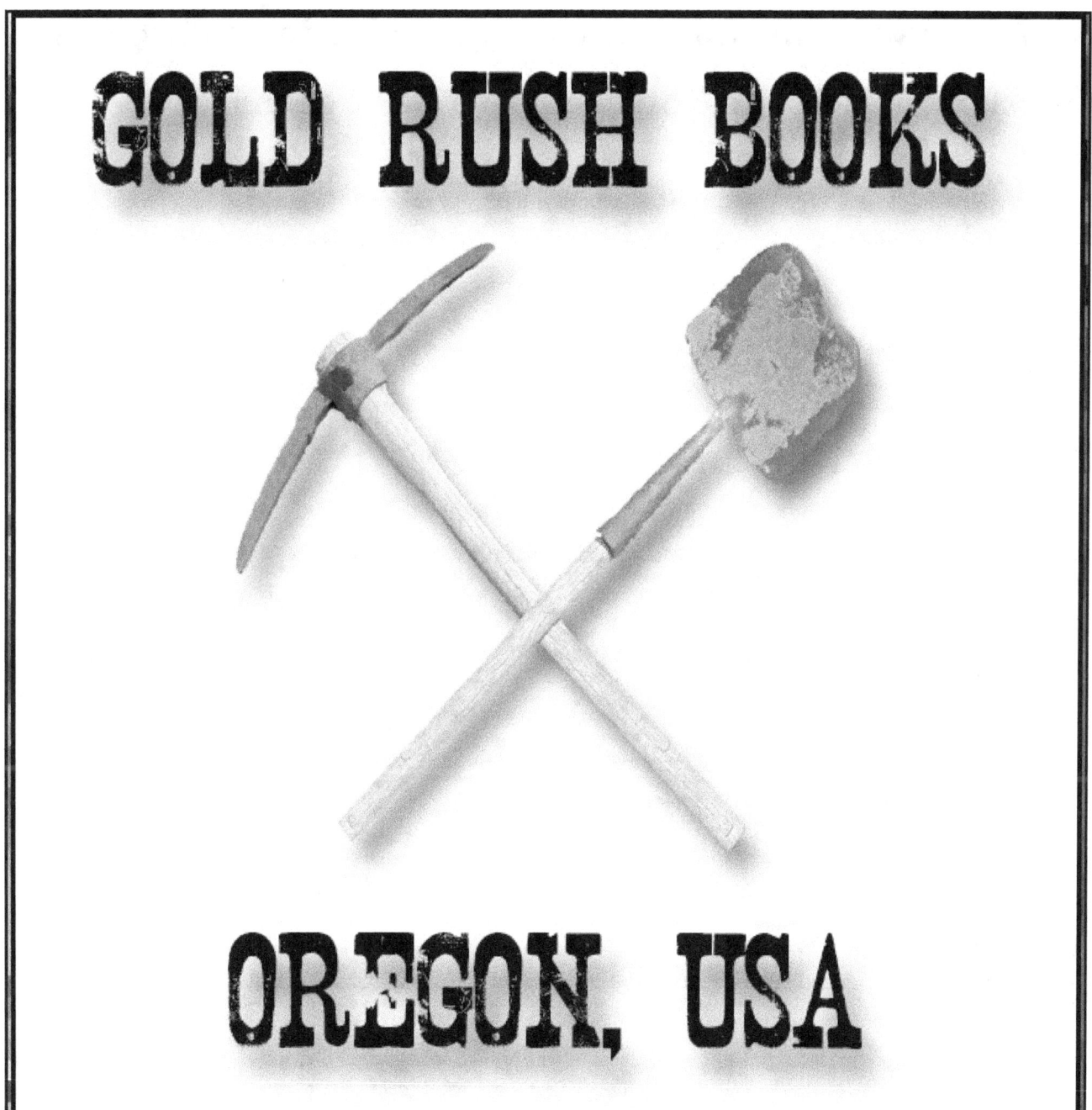

www.GoldMiningBooks.com

Books On Mining

Visit: www.goldminingbooks.com to order your copies or ask your favorite book seller to offer them.

Mining Books by Kerby Jackson

Gold Dust: Stories From Oregon's Mining Years - Oregon mining historian and prospector, Kerby Jackson, brings you a treasure trove of seventeen stories on Southern Oregon's rich history of gold prospecting, the prospectors and their discoveries, and the breathtaking areas they settled in and made homes. 5" X 8", 98 ppgs. Retail Price: $11.99

The Golden Trail: More Stories From Oregon's Mining Years - In his follow-up to "Gold Dust: Stories of Oregon's Mining Years", this time around, Jackson brings us twelve tales from Oregon's Gold Rush, including the story about the first gold strike on Canyon Creek in Grant County, about the old timers who found gold by the pail full at the Victor Mine near Galice, how Iradel Bray discovered a rich ledge of gold on the Coquille River during the height of the Rogue River War, a tale of two elderly miners on the hunt for a lost mine in the Cascade Mountains, details about the discovery of the famous Armstrong Nugget and others. 5" X 8", 70 ppgs. Retail Price: $10.99

Oregon Mining Books

Geology and Mineral Resources of Josephine County, Oregon - Unavailable since the 1970's, this important publication was originally compiled by the Oregon Department of Geology and Mineral Industries and includes important details on the economic geology and mineral resources of this important mining area in South Western Oregon. Included are notes on the history, geology and development of important mines, as well as insights into the mining of gold, copper, nickel, limestone, chromium and other minerals found in large quantities in Josephine County, Oregon. 8.5" X 11", 54 ppgs. Retail Price: $9.99

Mines and Prospects of the Mount Reuben Mining District - Unavailable since 1947, this important publication was originally compiled by geologist Elton Youngberg of the Oregon Department of Geology and Mineral Industries and includes detailed descriptions, histories and the geology of the Mount Reuben Mining District in Josephine County, Oregon. Included are notes on the history, geology, development and assay statistics, as well as underground maps of all the major mines and prospects in the vicinity of this much neglected mining district. 8.5" X 11", 48 ppgs. Retail Price: $9.99

The Granite Mining District - Notes on the history, geology and development of important mines in the well known Granite Mining District which is located in Grant County, Oregon. Some of the mines discussed include the Ajax, Blue Ribbon, Buffalo, Continental, Cougar-Independence, Magnolia, New York, Standard and the Tillicum. Also included are many rare maps pertaining to the mines in the area. 8.5" X 11", 48 ppgs. Retail Price: $9.99

Ore Deposits of the Takilma and Waldo Mining Districts of Josephine County, Oregon - The Waldo and Takilma mining districts are most notable for the fact that the earliest large scale mining of placer gold and copper in Oregon took place in these two areas. Included are details about some of the earliest large gold mines in the state such as the Llano de Oro, High Gravel, Cameron, Platerica, Deep Gravel and others, as well as copper mines such as the famous Queen of Bronze mine, the Waldo, Lily and Cowboy mines. This volume also includes six maps and 20 original illustrations. 8.5" X 11", 74 ppgs. Retail Price: $9.99

Metal Mines of Douglas, Coos and Curry Counties, Oregon - Oregon mining historian Kerby Jackson introduces us to a classic work on Oregon's mining history in this important re-issue of Bulletin 14C Volume 1, otherwise known as the Douglas, Coos & Curry Counties, Oregon Metal Mines Handbook. Unavailable since 1940, this important publication was originally compiled by the Oregon Department of Geology and Mineral Industries includes detailed descriptions, histories and the geology of over 250 metallic mineral mines and prospects in this rugged area of South West Oregon. 8.5" X 11", 158 ppgs. Retail Price: $19.99

Metal Mines of Jackson County, Oregon - Unavailable since 1943, this important publication was originally compiled by the Oregon Department of Geology and Mineral Industries includes detailed descriptions, histories and the geology of over 450 metallic mineral mines and prospects in Jackson County, Oregon. Included are such famous gold mining areas as Gold Hill, Jacksonville, Sterling and the Upper Applegate. **8.5" X 11", 220 ppgs. Retail Price: $24.99**

Metal Mines of Josephine County, Oregon - Oregon mining historian Kerby Jackson introduces us to a classic work on Oregon's mining history in this important re-issue of Bulletin 14C, otherwise known as the Josephine County, Oregon Metal Mines Handbook. Unavailable since 1952, this important publication was originally compiled by the Oregon Department of Geology and Mineral Industries includes detailed descriptions, histories and the geology of over 500 metallic mineral mines and prospects in Josephine County, Oregon. **8.5" X 11", 250 ppgs. Retail Price: $24.99**

Metal Mines of North East Oregon - Oregon mining historian Kerby Jackson introduces us to a classic work on Oregon's mining history in this important re-issue of Bulletin 14A and 14B, otherwise known as the North East Oregon Metal Mines Handbook. Unavailable since 1941, this important publication was originally compiled by the Oregon Department of Geology and Mineral Industries and includes detailed descriptions, histories and the geology of over 750 metallic mineral mines and prospects in North Eastern Oregon. **8.5" X 11", 310 ppgs. Retail Price: $29.99**

Metal Mines of North West Oregon - Oregon mining historian Kerby Jackson introduces us to a classic work on Oregon's mining history in this important re-issue of Bulletin 14D, otherwise known as the North West Oregon Metal Mines Handbook. Unavailable since 1951, this important publication was originally compiled by the Oregon Department of Geology and Mineral Industries and includes detailed descriptions, histories and the geology of over 250 metallic mineral mines and prospects in North Western Oregon. **8.5" X 11", 182 ppgs. Retail Price: $19.99**

Mines and Prospects of Oregon - Mining historian Kerby Jackson introduces us to a classic mining work by the Oregon Bureau of Mines in this important re-issue of The Handbook of Mines and Prospects of Oregon. Unavailable since 1916, this publication includes important insights into hundreds of gold, silver, copper, coal, limestone and other mines that operated in the State of Oregon around the turn of the 19th Century. Included are not only geological details on early mines throughout Oregon, but also insights into their history, production, locations and in some cases, also included are rare maps of their underground workings. **8.5" X 11", 314 ppgs. Retail Price: $24.99**

Lode Gold of the Klamath Mountains of Northern California and South West Oregon
(See California Mining Books)

Mineral Resources of South West Oregon - Unavailable since 1914, this publication includes important insights into dozens of mines that once operated in South West Oregon, including the famous gold fields of Josephine and Jackson Counties, as well as the Coal Mines of Coos County. Included are not only geological details on early mines throughout South West Oregon, but also insights into their history, production and locations. **8.5" X 11", 154 ppgs. Retail Price: $11.99**

Chromite Mining in The Klamath Mountains of California and Oregon
(See California Mining Books)

Southern Oregon Mineral Wealth - Unavailable since 1904, this rare publication provides a unique snapshot into the mines that were operating in the area at the time. Included are not only geological details on early mines throughout South West Oregon, but also insights into their history, production and locations. Some of the mining areas include Grave Creek, Greenback, Wolf Creek, Jump Off Joe Creek, Granite Hill, Galice, Mount Reuben, Gold Hill, Galls Creek, Kane Creek, Sardine Creek, Birdseye Creek, Evans Creek, Foots Creek, Jacksonville, Ashland, the Applegate River, Waldo, Kerby and the Illinois River, Althouse and Sucker Creek, as well as insights into local copper mining and other topics. **8.5" X 11", 64 ppgs. Retail Price: $8.99**

Geology and Ore Deposits of the Takilma and Waldo Mining Districts - Unavailable since the 1933, this publication was originally compiled by the United States Geological Survey and includes details on gold and copper mining in the Takilma and Waldo Districts of Josephine County, Oregon. The Waldo and Takilma mining districts are most notable for the fact that the earliest large scale mining of placer gold and copper in Oregon took place in these two areas. Included in this report are details about some of the earliest large gold mines in the state such as the Llano de Oro, High Gravel, Cameron, Platerica, Deep Gravel and others, as well as copper mines such as the famous Queen of Bronze mine, the Waldo, Lily and Cowboy mines. In addition to geological examinations, insights are also provided into the production, day to day operations and early histories of these mines, as well as calculations of known mineral reserves in the area. This volume also includes six maps and 20 original illustrations. **8.5" X 11", 74 ppgs. Retail Price: $9.99**

Gold Mines of Oregon - Oregon mining historian Kerby Jackson introduces us to a classic work on Oregon's mining history in this important re-issue of Bulletin 61, otherwise known as "Gold and Silver In Oregon". Unavailable since 1968, this important publication was originally compiled by geologists Howard C. Brooks and Len Ramp of the Oregon Department of Geology and Mineral Industries and includes detailed descriptions, histories and the geology of over 450 gold mines Oregon. Included are notes on the history, geology and gold production statistics of all the major mining areas in Oregon including the Klamath Mountains, the Blue Mountains and the North Cascades. While gold is where you find it, as every miner knows, the path to success is to prospect for gold where it was previously found. **8.5" X 11", 344 ppgs. Retail Price: $24.99**

Mines and Mineral Resources of Curry County Oregon - Originally published in 1916, this important publication on Oregon Mining has not been available for nearly a century. Included are rare insights into the history, production and locations of dozens of gold mines in Curry County, Oregon, as well as detailed information on important Oregon mining districts in that area such as those at Agness, Bald Face Creek, Mule Creek, Boulder Creek, China Diggings, Collier Creek, Elk River, Gold Beach, Rock Creek, Sixes River and elsewhere. Particular attention is especially paid to the famous beach gold deposits of this portion of the Oregon Coast. **8.5" X 11", 140 ppgs. Retail Price: $11.99**

Chromite Mining in South West Oregon - Originally published in 1961, this important publication on Oregon Mining has not been available for nearly a century. Included are rare insights into the history, production and locations of nearly 300 chromite mines in South Western Oregon. **8.5" X 11", 184 ppgs. Retail Price: $14.99**

Mineral Resources of Douglas County Oregon - Originally published in 1972, this important publication on Oregon Mining has not been available for nearly forty years. Included are rare insights into the geology, history, production and locations of numerous gold mines and other mining properties in Douglas County, Oregon. **8.5" X 11", 124 ppgs. Retail Price: $11.99**

Mineral Resources of Coos County Oregon - Originally published in 1972, this important publication on Oregon Mining has not been available for nearly forty years. Included are rare insights into the geology, history, production and locations of numerous gold mines and other mining properties in Coos County, Oregon. **8.5" X 11", 100 ppgs. Retail Price: $11.99**

Mineral Resources of Lane County Oregon - Originally published in 1938, this important publication on Oregon Mining has not been available for nearly seventy five years. Included are extremely rare insights into the geology and mines of Lane County, Oregon, in particular in the Bohemia, Blue River, Oakridge, Black Butte and Winberry Mining Districts. **8.5" X 11", 82 ppgs. Retail Price: $9.99**

Mineral Resources of the Upper Chetco River of Oregon: Including the Kalmiopsis Wilderness - Originally published in 1975, this important publication on Oregon Mining has not been available for nearly forty years. Withdrawn under the 1872 Mining Act since 1984, real insight into the minerals resources and mines of the Upper Chetco River has long been unavailable due to the remoteness of the area. Despite this, the decades of battle between property owners and environmental extremists over the last private mining inholding in the area has continued to pique the interest of those interested in mining and other forms of natural resource use. Gold mining began in the area in the 1850's and has a rich history in this geographic area, even if the facts surrounding it are little known. Included are twenty two rare photographs, as well as insights into the Becca and Morning Mine, the Emmly Mine (also known as Emily Camp), the Frazier Mine, the Golden Dream or Higgins Mine, Hustis Mine, Peck Mine and others. **8.5" X 11", 64 ppgs. Retail Price: $8.99**

Gold Dredging in Oregon - Originally published in 1939, this important publication on Oregon Mining has not been available for nearly seventy five years. Included are extremely rare insights into the history and day to day operations of the dragline and bucketline gold dredges that once worked the placer gold fields of South West and North East Oregon in decades gone by. Also included are details into the areas that were worked by gold dredges in Josephine, Jackson, Baker and Grant counties, as well as the economic factors that impacted this mining method. This volume also offers a unique look into the values of river bottom land in relation to both farming and mining, in how farm lands were mined, re-soiled and reclamated after the dredges worked them. Featured are hard to find maps of the gold dredge fields, as well as rare photographs from a bygone era. **8.5" X 11", 86 ppgs. Retail Price: $8.99**

Quick Silver Mining in Oregon - Originally published in 1963, this important publication on Oregon Mining has not been available for over fifty years. This publication includes details into the history and production of Elemental Mercury or Quicksilver in the State of Oregon. **8.5" X 11", 238 ppgs. Retail Price: $15.99**

Mines of the Greenhorn Mining District of Grant County Oregon - Originally published in 1948, this important publication on Oregon Mining has not been available for over sixty five years. In this publication are rare insights into the mines of the famous Greenhorn Mining District of Grant County, Oregon, especially the famous Morning Mine. Also included are details on the Tempest, Tiger, Bi-Metallic, Windsor, Psyche, Big Johnny, Snow Creek, Banzette and Paramount Mines, as well as prospects in the vicinities in the famous mining areas of Mormon Basin, Vinegar Basin and Desolation Creek. Included are hard to find mine maps and dozens of rare photographs from the bygone era of Grant County's rich mining history. **8.5" X 11", 72 ppgs. Retail Price: $9.99**

Geology of the Wallowa Mountains of Oregon: Part I (Volume 1) - Originally published in 1938, this important publication on Oregon Mining has not been available for nearly seventy five years. Included are details on the geology of this unique portion of North Eastern Oregon. This is the first part of a two book series on the area. Accompanying the text are rare photographs and historic maps.8.5" X 11", 92 ppgs. **Retail Price: $9.99**

Geology of the Wallowa Mountains of Oregon: Part II (Volume 2) - Originally published in 1938, this important publication on Oregon Mining has not been available for nearly seventy five years. Included are details on the geology of this unique portion of North Eastern Oregon. This is the first part of a two book series on the area. Accompanying the text are rare photographs and historic maps.8.5" X 11", 94 ppgs. **Retail Price: $9.99**

Field Identification of Minerals For Oregon Prospectors - Originally published in 1940, this important publication on Oregon Mining has not been available for nearly seventy five years. Included in this volume is an easy system for testing and identifying a wide range of minerals that might be found by prospectors, geologists and rockhounds in the State of Oregon, as well as in other locales. Topics include how to put together your own field testing kit and how to conduct rudimentary tests in the field. This volume is written in a clear and concise way to make it useful even for beginners. **8.5" X 11", 158 ppgs. Retail Price: $14.99**

The Bohemia Mining District of Oregon - Originally published in 1900, this important publication on Oregon Mining has not been available for over a century. Included in this volume are important insights into the famous Bohemia Mining District of Oregon, including the histories and locations of important gold mines in the area such as the Ophir Mine, Clarence, Acturas, Peek-a-boo, White Swan, Combination Mine, the Musick Mine, The California, White Ghost, The Mystery, Wall Street, Vesuvius, Story, Lizzie Bullock, Delta, Elsie Dora, Golden Slipper, Broadway, Champion Mine, Knott, Noonday, Helena, White Wings, Riverside and others. Also included are notes on the nearby Blue River Mining District. **8.5" X 11", 58 ppgs. Retail Price: $9.99**

The Gold Fields of Eastern Oregon - Unavailable since 1900, this publication was originally compiled by the Baker City Chamber of Commerce Offering important insights into the gold mining history of Eastern Oregon, "The Gold Fields of Eastern Oregon" sheds a rare light on many of the gold mines that were operating at the turn of the 19th Century in Baker County and Grant County in North Eastern Oregon. Some of the areas featured include the Cable Cove District, Baisely-Elhorn, Granite, Red Boy, Bonanza, Susanville, Sparta, Virtue, Vaughn, Sumpter, Burnt River, Rye Valley and other mining districts. Included is basic information on not only many gold mines that are well known to those interested in Eastern Oregon mining history, but also many mines and prospects which have been mostly lost to the passage of time. Accompanying are numerous rare photos **8.5" X 11", 78 ppgs. Retail Price: $10.99**

Gold Mining in Eastern Oregon - Originally published in 1938, this important publication on Oregon Mining has not been available for over a century. Included in this volume are important insights into the famous mining districts of Eastern Oregon during the late 1930's. Particular attention is given to those gold mines with milling and concentrating facilities in the Greenhorn, Red Boy, Alamo, Bonanza, Granite, Cable Cove, Cracker Creek, Virtue, Keating, Medical Springs, Sanger, Sparta, Chicken Creek, Mormon Basin, Connor Creek, Cornucopia and the Bull Run Mining Districts. Some of the mines featured include the Ben Harrison, North Pole-Columbia, Highland Maxwell, Baisley-Elkhorn, White Swan, Balm Creek, Twin Baby, Gem of Sparta, New Deal, Gleason, Gifford-Johnson, Cornucopia, Record, Bull Run, Orion and others. Of particular interest are the mill flow sheets and descriptions of milling operations of these mines. **8.5" X 11", 68 ppgs. Retail Price: $8.99**

The Gold Belt of the Blue Mountains of Oregon - Originally published in 1901, this important publication on Oregon Mining has not been available for over a century. Included in this volume are rare insights into the gold deposits of the Blue Mountains of North East Oregon, including the history of their early discovery and early production. Extensive details are offered on this important mining area's mineralogy and economic geology, as well as insights into nearby gold placers, silver deposits and copper deposits. Featured are the Elkhorn and Rock Creek mining districts, the Pocahontas district, Auburn and Minersville districts, Sumpter and Cracker Creek, Cable Cove, the Camp Carson district, Granite, Alamo, Greenhorn, Robinsonville, the Upper Burnt River Valley and Bonanza districts, Susanville, Quartzburg, Canyon Creek, Virtue, the Copper Butte district, the North Powder River, Sparta, Eagle Creek, Cornucopia, Pine Creek, Lower Powder River, the Upper Snake River Canyon, Rye Valley, Lower Burnt River Valley, Mormon Basin, the Malheur and Clarks Creek districts, Sutton Creek and others. Of particular interest are important details on numerous gold mines and prospects in these mining districts, including their locations, histories, geology and other important information, as well as information on silver, copper and fire opal deposits. **8.5" X 11", 250 ppgs. Retail Price: $24.99**

<u>Mining in the Cascades Range of Oregon</u> - Originally published in 1938, this important publication on Oregon Mining has not been available for over seventy five years. Included in this volume are rare insights into the gold mines and other types of metal mines in the Cascades Mountain Range of Oregon. Some of the important mining areas covered include the famous Bohemia Mining District, the North Santiam Mining District, Quartzville Mining District, Blue River Mining District, Fall Creek Mining District, Oakridge District, Zinc District, Buzzard-Al Sarena District, Grand Cove, Climax District and Barron Mining District. Of particular interest are important details on over 100 mines and prospects in these mining districts, including their locations, histories, geology and other important information. **8.5" X 11", 170 ppgs. Retail Price: $14.99**

<u>Beach Gold Placers of the Oregon Coast</u> - Originally published in 1934, this important publication on Oregon Mining has not been available for over 80 years. Included in this volume are rare insights into the beach gold deposits of the State of Oregon, including their locations, occurance, composition and geology. Of particular interest is information on placer platinum in Oregon's rich beach deposits. Also included are the locations and other information on some famous Oregon beach mines, including the Pioneer, Eagle, Chickamin, Iowa and beach placer mines north of the mouth of the Rogue River. **8.5" X 11", 60 ppgs. Retail Price: $8.99**

Idaho Mining Books

<u>Gold in Idaho</u> - Unavailable since the 1940's, this publication was originally compiled by the Idaho Bureau of Mines and includes details on gold mining in Idaho. Included is not only raw data on gold production in Idaho, but also valuable insight into where gold may be found in Idaho, as well as practical information on the gold bearing rocks and other geological features that will assist those looking for placer and lode gold in the State of Idaho. This volume also includes thirteen gold maps that greatly enhance the practical usability of the information contained in this small book detailing where to find gold in Idaho. **8.5" X 11", 72 ppgs. Retail Price: $9.99**

<u>Geology of the Couer D'Alene Mining District of Idaho</u> - Unavailable since 1961, this publication was originally compiled by the Idaho Bureau of Mines and Geology and includes details on the mining of gold, silver and other minerals in the famous Coeur D'Alene Mining District in Northern Idaho. Included are details on the early history of the Coeur D'Alene Mining District, local tectonic settings, ore deposit features, information on the mineral belts of the Osburn Fault, as well as detailed information on the famous Bunker Hill Mine, the Dayrock Mine, Galena Mine, Lucky Friday Mine and the infamous Sunshine Mine. This volume also includes sixteen hard to find maps. **8.5" X 11", 70 ppgs. Retail Price: $9.99**

<u>The Gold Camps and Silver Cities of Idaho</u> - Originally published in 1963, this important publication on Idaho Mining has not been available for nearly fifty years. Included are rare insights into the history of Idaho's Gold Rush, as well as the mad craze for silver in the Idaho Panhandle. Documented in fine detail are the early mining excitements at Boise Basin, at South Boise, in the Owyhees, at Deadwood, Long Valley, Stanley Basin and Robinson Bar, at Atlanta, on the famous Boise River, Volcano, Little Smokey, Banner, Boise Ridge, Hailey, Leesburg, Lemhi, Pearl, at South Mountain, Shoup and Ulysses, Yellow Jacket and Loon Creek. The story follows with the appearance of Chinese miners at the new mining camps on the Snake River, Black Pine, Yankee Fork, Bay Horse, Clayton, Heath, Seven Devils, Gibbonsville, Vienna and Sawtooth City. Also included are special sections on the Idaho Lead and Silver mines of the late 1800's, as well as the mining discoveries of the early 1900's that paved the way for Idaho's modern mining and mineral industry. Lavishly illustrated with rare historic photos, this volume provides a one of a kind documentary into Idaho's mining history that is sure to be enjoyed by not only modern miners and prospectors who still scour the hills in search of nature's treasures, but also those enjoy history and tromping through overgrown ghost towns and long abandoned mining camps. **8.5" X 11", 186 ppgs. Retail Price: $14.99**

<u>Ore Deposits and Mining in North Western Custer County Idaho</u> - Unavailable since 1913, this important publication was originally published by the Us Department of the Interior and has been unavailable for a century. Included are fine details on the geology, geography, gold placers and gold and silver bearing quartz veins of the mining region of North West Custer County, Idaho. Of particular interest is a rare look at the mines and prospects of the region, including those such as the Ramshorn Mine, SkyLark, Riverview, Excelsior, Beardsley, Pacific, Hoosier, Silver Brick, Forest Rose and dozens of others in the Bay Horse Mining District. Also covered are the mines of the Yankee Fork District such as the Lucky Boy, Badger, Black, Enterprise, Charles Dickens, Morrison, Golden Sunbeam, Montana, Golden Gate and others, as well as those in the Loon Mining District. **8.5" X 11", 126 ppgs. Retail Price: $12.99**

Gold Rush To Idaho - Unavailable since 1963, this important publication was originally published by the Idaho Bureau of Mines and has been unavailable for 50 years. "Gold Rush To Idaho" revisits the earliest years of the discovery of gold in Idaho Territory and introduces us to the conditions that the pioneer gold seekers met when they blazed a trail through the wilderness of Idaho's mountains and discovered the precious yellow metal at Oro Fino and Pierce. Subsequent rushes followed at places like Elk City, Newsome, Clearwater Station, Florence, Warrens and elsewhere. Of particular interest is a rare look at the hardships that the first miners in Idaho met with during their day to day existences and their attempts to bring law and order to their mining camps. **8.5" X 11", 88 ppgs. Retail Price: $9.99**

The Geology and Mines of Northern Idaho and North Western Montana - Unavailable since 1909, this important publication was originally published by the Us Department of the Interior and has been unavailable for a century. Included are fine details on the geology and geography of the mining regions of Northern Idaho and North Western Montana. Of particular interest is a rare look at the mines and prospects of the region, including those in the Pine Creek Mining District, Lake Pend Oreille district, Troy Mining District, Sylvanite District, Cabinet Mining District, Prospect Mining District and the Missoula Valley. Some of the mines featured include the Iron Mountain, Silver Butte, Snowshoe, Grouse Mountain Mine and others. **8.5" X 11", 142 ppgs. Retail Price: $12.99**

Mining in the Alturas Quadrangle of Blaine County Idaho - Unavailable since 1922, this important publication was originally published by the Idaho Bureau of Mines and has been unavailable for ninety years. Topics include the geology, rock formations and the formation of ore deposits in this important mining area of Idaho. Of particular focus is information on the local geology, quartz veins and ore deposits of this portion of Idaho. Included are hard to find details, including the descriptions and locations of numerous gold and silver mines in the area including the Silver King, Pilgrim, Columbia, Lone Jack, Sunbeam, Pride of the West, Lucky Boy, Scotia, Atlanta, Beaver-Bidwell and others mines and prospects. **8.5" X 11", 56 ppgs. Retail Price: $8.99**

Mining in Lemhi County Idaho - Originally published in 1913, this important book on Idaho Mining has not been available to miners for over a century. Included are rare insights into hundreds of gold, silver, copper and other mines in this famous Idaho mining area. Details include the locations, geology, history, production and other facts of the mines of this region, not only gold and silver hardrock mines, but also gold placer mines, lead-silver deposits, copper mines, cobalt-nickel deposits, tungsten and tin mines . It is lavishly illustrated with hard to find photos of the period and rare mining maps. Some of the vicinities featured include the Nicholia Mining District, Spring Mountain District, Texas District, Blue Wing District, Junction District, McDevitt District, Pratt Creek, Eldorado District, Kirtley Creek, Carmen Creek, Gibbonsville, Indian Creek, Mineral Hill District, Mackinaw, Eureka District, Blackbird District, YellowJacket District, Gravel Range District, Junction District, Parker Mountain and other mining districts. **8.5" X 11", 226 ppgs. Retail Price: $19.99**

Utah Mining Books

Fluorite in Utah - Unavailable since 1954, this publication was originally compiled by the USGS, State of Utah and U.S. Atomic Energy Commission and details the mining of fluorspar, also known as fluorite in the State of Utah. Included are details on the geology and history of fluorspar (fluorite) mining in Utah, including details on where this unique gem mineral may be found in the State of Utah. **8.5" X 11", 60 ppgs. Retail Price: $8.99**

California Mining Books

The Tertiary Gravels of the Sierra Nevada of California - Mining historian Kerby Jackson introduces us to a classic mining work by Waldemar Lindgren in this important re-issue of The Tertiary Gravels of the Sierra Nevada of California. Unavailable since 1911, this publication includes details on the gold bearing ancient river channels of the famous Sierra Nevada region of California. **8.5" X 11", 282 ppgs. Retail Price: $19.99**

The Mother Lode Mining Region of California - Unavailable since 1900, this publication includes details on the gold mines of California's famous Mother Lode gold mining area. Included are details on the geology, history and important gold mines of the region, as well as insights into historic mining methods, mine timbering, mining machinery, mining bell signals and other details on how these mines operated. Also included are insights into the gold mines of the California Mother Lode that were in operation during the first sixty years of California's mining history. **8.5" X 11", 176 ppgs. Retail Price: $14.99**

Lode Gold of the Klamath Mountains of Northern California and South West Oregon - Unavailable since 1971, this publication was originally compiled by Preston E. Hotz and includes details on the lode mining districts of Oregon and California's Klamath Mountains. Included are details on the geology, history and important lode mines of the French Gulch, Deadwood, Whiskeytown, Shasta, Redding, Muletown, South Fork, Old Diggings, Dog Creek (Delta), Bully Choop (Indian Creek), Harrison Gulch, Hayfork, Minersville, Trinity Center, Canyon Creek, East Fork, New River, Denny, Liberty (Black Bear), Cecilville, Callahan, Yreka, Fort Jones and Happy Camp mining districts in California, as well as the Ashland, Rogue River, Applegate, Illinois River, Takilma, Greenback, Galice, Silver Peak, Myrtle Creek and Mule Creek districts of South Western Oregon. Also included are insights into the mineralization and other characteristics of this important mining region. **8.5" X 11", 100 ppgs. Retail Price: $10.99**

Mines and Mineral Resources of Shasta County, Siskiyou County, Trinity County: California - Unavailable since 1915, this publication was originally compiled by the California State Mining Bureau and includes details on the gold mines of this area of Northern California. Also included are insights into the mineralization and other characteristics of this important mining region, as well as the location of historic gold mines. **8.5" X 11", 204 ppgs. Retail Price: $19.99**

Geology of the Yreka Quadrangle, Siskiyou County, California - Unavailable since 1977, this publication was originally compiled by Preston E. Hotz and includes details on the geology of the Yreka Quadrangle of Siskiyou County, California. Also included are insights into the mineralization and other characteristics of this important mining region. **8.5" X 11", 78 ppgs. Retail Price: $7.99**

Mines of San Diego and Imperial Counties, California - Originally published in 1914, this important publication on California Mining has not been available for a century. This publication includes important information on the early gold mines of San Diego and Imperial County, which were some of the first gold fields mined in California by early Spanish and Mexican miners before the 49ers came on the scene. Included are not only details on early mining methods in the area, production statistics and geological information, but also the location of the early gold mines that helped make California "The Golden State". Also included are details on the mining of other minerals such as silver, lead, zinc, manganese, tungsten, vanadium, asbestos, barite, borax, cement, clay, dolomite, fluospar, gem stones, graphite, marble, salines, petroleum, stronium, talc and others. **8.5" X 11", 116 ppgs. Retail Price: $12.99**

Mines of Sierra County, California - Unavailable since 1920, this publication was originally compiled by the California State Mining Bureau and includes details on the gold mines of Sierra County, California. Also included are insights into the mineralization and other characteristics of this important mining region, as well as the location of historic gold mines. **8.5" X 11", 156 ppgs. Retail Price: $19.99**

Mines of Plumas County, California - Unavailable since 1918, this publication was originally compiled by the California State Mining Bureau and includes details on the gold mines of Plumas County, California. Also included are insights into the mineralization and other characteristics of this important mining region, as well as the location of historic gold mines. **8.5" X 11", 200 ppgs. Retail Price: $19.99**

Mines of El Dorado, Placer, Sacramento and Yuba Counties, California - Originally published in 1917, this important publication on California Mining has not been available for nearly a century. This publication includes important information on the early gold mines of El Dorado County, Placer County, Sacramento County and Yuba County, which were some of the first gold fields mined by the Forty-Niners during the California Gold Rush. Included are not only details on early mining methods in the area, production statistics and geological information, but also the location of the early gold mines that helped make California "The Golden State". Also included are insights into the early mining of chrome, copper and other minerals in this important mining area. **8.5" X 11", 204 ppgs. Retail Price: $19.99**

Mines of Los Angeles, Orange and Riverside Counties, California - Originally published in 1917, this important publication on California Mining has not been available for nearly a century. This publication includes important information on the early gold mines of Los Angeles County, Orange County and Riverside County, which were some of the first gold fields mined in California by early Spanish and Mexican miners before the 49ers came on the scene. Included are not only details on early mining methods in the area, production statistics and geological information, but also the location of the early gold mines that helped make California "The Golden State". **8.5" X 11", 146 ppgs. Retail Price: $12.99**

Mines of San Bernadino and Tulare Counties, California - Originally published in 1917, this important publication on California Mining has not been available for nearly a century. This publication includes important information on the early gold mines of San Bernadino and Tulare County, which were some of the first gold fields mined in California by early Spanish and Mexican miners before the 49ers came on the scene. Included are not only details on early mining methods in the area, production statistics and geological information, but also the location of the early gold mines that helped make California "The Golden State". Also included are details on the mining of other minerals such as copper, iron, lead, zinc, manganese, tungsten, vanadium, asbestos, barite, borax, cement, clay, dolomite, fluospar, gem stones, graphite, marble, salines, petroleum, stronium, talc and others. **8.5" X 11", 200 ppgs. Retail Price: $19.99**

Chromite Mining in The Klamath Mountains of California and Oregon - Unavailable since 1919, this publication was originally compiled by J.S. Diller of the United States Department of Geological Survey and includes details on the chromite mines of this area of Northern California and Southern Oregon. Also included are insights into the mineralization and other characteristics of this important mining region, as well as the location of historic mines. Also included are insights into chromite mining in Eastern Oregon and Montana. **8.5" X 11", 98 ppgs. Retail Price: $9.99**

Mines and Mining in Amador, Calaveras and Tuolumne Counties, California - Unavailable since 1915, this publication was originally compiled by William Tucker and includes details on the mines and mineral resources of this important California mining area. Included are details on the geology, history and important gold mines of the region, as well as insights into other local mineral resources such as asbestos, clay, copper, talc, limestone and others. Also included are insights into the mineralization and other characteristics of this important portion of California's Mother Lode mining region. 8.5" X 11", 198 ppgs. Retail Price: $14.99

The Cerro Gordo Mining District of Inyo County California - Unavailable since 1963, this publication was originally compiled by the United States Department of Interior. Included are insights into the mineralization and other characteristics of this important mining region of Southern California. Topics include the mining of gold and silver in this important mining district in Inyo County, California, including details on the history, production and locations of the Cerro Gordo Mine, the Morning Star Mine, Estelle Tunnel, Charles Lease Tunnel, Ignacio, Hart, Crosscut Tunnel, Sunset, Upper Newtown, Newtown, Ella, Perseverance, Newsboy, Belmont and other silver and gold mines in the Cerro Gordo Mining District. This volume also includes important insights into the fossil record, geologic formations, faults and other aspects of economic geology in this California mining district. 8.5" X 11", 104 ppgs. Retail Price: $10.99

Mining in Butte, Lassen, Modoc, Sutter and Tehama Counties of California - Unavailable since 1917, this publication was originally compiled by the United States Department of Interior. Included are insights into the mineralization and other characteristics of this important mining region of California. Topics include the mining of asbestos, chromite, gold, diamonds and manganese in Butte County, the mining of gold and copper in the Hayden Hill and Diamond Mountain mining districts of Lassen County, the mining of coal, salt, copper and gold in the High Grade and Winters mining districts of Modoc County, gold mining in Sutter County and the mining of gold, chromite, manganese and copper in Tehama County. This volume also includes the production records and locations of numerous mines in this important mining region. 8.5" X 11", 114 ppgs. Retail Price: $11.99

Mines of Trinity County California - Originally published in 1965, this important publication on California Mining has not been available for nearly fifty years. This publication includes important information on mines and mining in Trinity County, California, as well insights into the mineralization and geology of this important mining area in Northern California. Included are extensive details on hardrock and placer gold mines and prospects, including charts showing the locations of these historic mines.. 8.5" X 11", 144 ppgs. Retail Price: $12.99

Mines of Kern County California - Originally published in 1962, this important publication on California Mining has not been available for nearly fifty years. This publication includes important information on mines and mining in Kern County, California, as well insights into the mineralization and geology of this important mining area in California. Included are extensive details on hardrock and placer gold mines and prospects, including charts showing the locations of these historic mines. 8.5" X 11", 398 ppgs. Retail Price: $24.99

Mines of Calaveras County California - Originally published in 1962, this important publication on California Mining has not been available for nearly fifty years. This publication includes important information on mines and mining in Calaveras County, California, as well insights into the mineralization and geology of this important mining area in Northern California. Included are extensive details on hardrock and placer gold mines and prospects, including charts showing the locations of these historic mines. 8.5" X 11", 236 ppgs. Retail Price: $19.99

Lode Gold Mining in Grass Valley California - Unavailable since 1940, this publication was originally compiled by the United States Department of Interior. Included are insights into the gold mineralization and other characteristics of this important mining region of Nevada County, California. This volume also includes important insights into the geologic formations, faults and other aspects of economic geology in this California mining district. Of particular interest are the fine details on many hardrock gold mines in the area, including their locations, histories, development and mineralization. Some of the mines featured include the Gold Hill Mine, Massachusetts Hill, Boundary, Peabody, Golden Center, North Star, Omaha, Lone Jack, Homeward Bound, Hartery, Wisconsin, Allison Ranch, Phoenix, Kate Hayes, W.Y.O.D., Empire, Rich Hill, Daisy Hill, Orleans, Sultana, Centennial, Conlin, Ben Franklin, Crown Point and many others. 8.5" X 11", 148 ppgs. Retail Price: $12.99

Lode Mining in the Alleghany District of Sierra County California - Unavailable since 1913, this publication was originally compiled by the United States Department of Interior. Included are insights into the mineralization and other characteristics of this important mining region of Sierra County. Included are details on the history, production and locations of numerous hardrock gold mines in this famous California area, including the Tightner Mine, Minnie D., Osceola, Eldorado, Twenty One, Sherman, Kenton, Oriental, Rainbow, Plumbago, Irelan, Gold Canyon, North Fork, Federal, Kate Hardy and others. This volume also includes important insights into the fossil record, geologic formations, faults and other aspects of economic geology in this California mining district. 8.5" X 11", 48 ppgs. Retail Price: $7.99

<u>Six Months In The Gold Mines During The California Gold Rush</u> - Unavailable since 1850, this important work is a first hand account of one "49'ers" personal experience during the great California Gold Rush, shedding important light on one of the most exciting periods in the history of not only California, but also the world. Compiled from journals written between 1847 and 1849 by E. Gould Buffum, a native of New York, "Six Months In The Gold Mines During The California Gold Rush" offers a rare look into the day to day lives of the people who came to California to work in her gold mines when the state was still a great frontier. **8.5" X 11", 290 ppgs. Retail Price: $19.99**

<u>Quartz Mines of the Grass Valley Mining District of California</u> - Unavailable since 1867, this important publication has not been available since those days. This rare publication offers a short dissertation on the early hardrock mines in this important mining district in the California Mother Lode region between the 1850's and 1860's. Also included are hard to find details on the mineralization and locations of these mines, as well as how they were operated in those day. **8.5" X 11", 44 ppgs. Retail Price: $8.99**

Alaska Mining Books

<u>Ore Deposits of the Willow Creek Mining District, Alaska</u> - Unavailable since 1954, this hard to find publication includes valuable insights into the Willow Creek Mining District near Hatcher Pass in Alaska. The publication includes insights into the history, geology and locations of the well known mines in the area, including the Gold Cord, Independence, Fern, Mabel, Lonesome, Snowbird, Schroff-O'Neil, High Grade, Marion Twin, Thorpe, Webfoot, Kelly-Willow, Lane, Holland and others. **8.5" X 11", 96 ppgs. Retail Price: $9.99**

<u>The Juneau Gold Belt of Alaska</u> - Unavailable since 1906, this hard to find publication includes valuable insights into the gold mines around Juneau, Alaska. The publication includes important details into the history, geology and locations of the well known gold mines and prospects in the area, including those around Windham Bay, Holkham Bay, Port Snettisham, on Grindstone and Rhine Creeks, Gold Creek, Douglas Island, Salmon Creek, Lemon Creek, Nugget Creek, from the Mendenhall River to Berners Bay, McGinnis Creek, Montana Creek, Peterson Creek, Windfall Creek, the Eagle River, Yankee Basin, Yankee Curve, Kowee Creek and elsewhere. Not only are gold placer mines included, but also hardrock gold mines. **8.5" X 11", 224 ppgs. Retail Price: $19.99**

Arizona Mining Books

<u>Mines and Mining in Northern Yuma County Arizona</u> - Originally published in 1911, this important publication on Arizona Mining has not been available for over a hundred years. Included are rare insights into the gold, silver, copper and quicksilver mines of Yuma County, Arizona together with hard to find maps and photographs. Some of the mines and mining districts featured include the Planet Copper Mine, Mineral Hill, the Clara Consolidated Mine, Viati Mine, Copper Basin prospect, Bowman Mine, Quartz King, Billy Mack, Carnation, the Wardwell and Osbourne, Valensuella Copper, the Mariquita, Colonial Mine, the French American, the New York-Plomosa, Guadalupe, Lead Camp, Mudersbach Copper Camp, Yellow Bird, the Arizona Northern (Salome Strike), Bonanza (Harqua Hala), Golden Eagle, Hercules, Socorro and others. **8.5" X 11", 144 ppgs. Retail Price: $11.99**

<u>The Aravaipa and Stanley Mining Districts of Graham County Arizona</u> - Originally published in 1925, this important publication on Arizona Mining has not been available for nearly ninety years. Included are rare insights into the gold and silver mines of these two important mining districts, together with hard to find maps. **8.5" X 11", 140 ppgs. Retail Price: $11.99**

<u>Gold in the Gold Basin and Lost Basin Mining Districts of Mohave County, Arizona</u> - This volume contains rare insights into the geology and gold mineralization of the Gold Basin and Lost Basin Mining Districts of Mohave County, Arizona that will be of benefit to miners and prospectors. Also included is a significant body of information on the gold mines and prospects of this portion of Arizona. This volume is lavishly illustrated with rare photos and mining maps. **8.5" X 11", 188 ppgs. Retail Price: $19.99**

<u>Mines of the Jerome and Bradshaw Mountains of Arizona</u> - This important publication on Arizona Mining has not been available for ninety years. This volume contains rare insights into the geology and ore deposits of the Jerome and Bradshaw Mountains of Arizona that will be of benefit to miners and prospectors who work those areas. Included is a significant body of information on the mines and prospects of the Verde, Black Hills, Cherry Creek, Prescott, Walker, Groom Creek, Hassayampa, Bigbug, Turkey Creek, Agua Fria, Black Canyon, Peck, Tiger, Pine Grove, Bradshaw, Tintop, Humbug and Castle Creek Mining Districts. This volume is lavishly illustrated with rare photos and mining maps. **8.5" X 11", 218 ppgs. Retail Price: $19.99**

<u>The Ajo Mining District of Pima County Arizona</u> - This important publication on Arizona Mining has not been available for nearly seventy years. This volume contains rare insights into the geology and mineralization of the Ajo Mining District in Pima County, Arizona and in particular the famous New Cornelia Mine. **8.5" X 11", 126 ppgs. Retail Price: $11.99**

<u>Mining in the Santa Rita and Patagonia Mountains of Arizona</u> - Originally published in 1915, this important publication on Arizona Mining has not been available for nearly a century. Included are rare insights into hundreds of gold, silver, copper and other mines in this famous Arizona mining area. Details include the locations, geology, history, production and other facts of the mines of this region. **8.5" X 11", 394 ppgs. Retail Price: $24.99**

<u>Mining in the Bisbee Quadrangle of Arizona</u> - Originally published in 1906, this important publication on Arizona Mining has not been available for nearly a century. Included are rare insights into hundreds of gold, silver, copper and other mines in this famous Arizona mining area. Details include the locations, geology, history, production and other facts of the mines of this important mining region. **8.5" X 11", 188 ppgs. Retail Price: $14.99**

Montana Mining Books

<u>A History of Butte Montana: The World's Greatest Mining Camp</u> - First published in 1900 by H.C. Freeman, this important publication sheds a bright light on one of the most important mining areas in the history of The West. Together with his insights, as well as rare photographs of the periods, Harry Freeman describes Butte and its vicinity from its early beginnings, right up to its flush years when copper flowed from its mines like a river. At the time of publication, Butte, Montana was known worldwide as "The Richest Mining Spot On Earth" and produced not only vast amounts of copper, but also silver, gold and other metals from its mines. Freeman illustrates, with great detail, the most important mines in the vicinity of Butte, providing rare details on their owners, their history and most importantly, how the mines operated and how their treasures were extracted. Of particular interest are the dozens of rare photographs that depict mines such as the famous Anaconda, the Silver Bow, the Smoke House, Moose, Paulin, Buffalo, Little Minah, the Mountain Consolidated, West Greyrock, Cora, the Green Mountain, Diamond, Bell, Parnell, the Neversweat, Nipper, Original and many others.   **8.5" X 11", 142 ppgs. Retail Price: $12.99**

<u>The Butte Mining District of Montana</u> - This important publication on Montana Mining has not been available for over a century. Included are rare insights into the gold, copper and silver mines of Butte, Montana together with hard to find maps and photographs. Some of the topics include the early history of gold, silver and copper mining in the Butte area, insight into the geology of its mining areas, the local distribution of gold, silver and copper ores, as well their composition and how to identify them. Also included are detailed facts about the mines in the Butte Mining District, including the famous Anaconda Mine, Gagnon, Parrot, Blue Vein, Moscow, Poulin, Stella, Buffalo, Green Mountain, Wake Up Jim, the Diamond-Bell Group, Mountain Consolidated, East Greyrock, West Greyrock, Snowball, Corra, Speculator, Adirondack, Miners Union, the Jessie-Edith May Group, Otisco, Iduna, Colorado, Lizzie, Cambers, Anderson, Hesperus, Preferencia and dozens of others. **8.5" X 11", 298 ppgs. Retail Price: $24.99**

<u>Mines of the Helena Mining Region of Montana</u> - This important publication on Montana Mining has not been available for over a century. Included are rare insights into the gold, copper and silver mines of the vicinity of Helena, Montana, including the Marysville Mining District, Elliston Mining District, Rimini Mining District, Helena Mining District, Clancy Mining District, Wickes Mining District, Boulder and Basin Mining Districts and the Elkhorn Mining District. Some of the topics include the early history of gold, silver and copper mining in the Helena area, insight into the geology of its mining areas, the local distribution of gold, silver and copper ores, as well their composition and how to identify them. Also included are detailed facts, history, geology and locations of over one hundred gold, silver and copper mines in the area . **8.5" X 11", 162 ppgs, Retail Price: $14.99**

<u>Mines and Geology of the Garnet Range of Montana</u> - This important publication on Montana Mining has not been available for over a century. Included are rare insights into the gold, copper and silver mines of the vicinity of this important mining area of Montana. Some of the topics include the early history of gold, silver and copper mining in the Garnet Mountains, insight into the geology of its mining areas, the local distribution of gold, silver and copper ores, as well their composition and how to identify them. Also included are detailed facts, history, geology and locations of numerous gold, silver and copper mines in the area . **8.5" X 11", 100 ppgs, Retail Price: $11.99**

<u>Mines and Geology of the Philipsburg Quadrangle of Montana</u> - This important publication on Montana Mining has not been available for over a century. Included are rare insights into the gold, copper and silver mines of the vicinity of this important mining area of Montana. Some of the topics include the early history of gold, silver and copper mining in the Philipsburg Quadrangle, insight into the geology of its mining areas, the local distribution of gold, silver and copper ores, as well their composition and how to identify them. Also included are detailed facts, history, geology and locations of over one hundred gold, silver and copper mines in the area **8.5" X 11", 290 ppgs, Retail Price: $24.99**

<u>Geology of the Marysville Mining District of Montana</u> - Included are rare insights into the mining geology of the Marysville Mining District. Some of the topics include the early history of gold, silver and copper mining in the area, insight into the geology of its mining areas, the local distribution of gold, silver and copper ores, as well their composition and how to identify them. Also included are detailed facts, history, geology and locations of gold, silver and copper mines in the area **8.5" X 11", 198 ppgs, Retail Price: $19.99**

<u>The Geology and Mines of Northern Idaho and North Western Montana</u>

See listing under Idaho.

Nevada Mining Books

<u>The Bull Frog Mining District of Nevada</u> - Unavailable since 1910, this publication was originally compiled by the United States Department of Interior. This volume also includes important insights into the geologic formations, faults and other aspects of economic geology in this Nevada mining district. Of particular interest are the fine details on many mines in the area, including their locations, histories, development and mineralization. Some of the mines featured include the National Bank Mine, Providence, Gibraltor, Tramps, Denver, Original Bullfrog, Gold Bar, Mayflower, Homestake-King and other mines and prospects. **8.5" X 11", 152 ppgs, Retail Price: $14.99**

<u>History of the Comstock Lode</u> - Unavailable since 1876, this publication was originally released by John Wiley & Sons. This volume also includes important insights into the famous Comstock Lode of Nevada that represented the first major silver discovery in the United States. During its spectacular run, the Comstock produced over 192 million ounces of silver and 8.2 million ounces of gold. Not only did the Comstock result in one of the largest mining rushes in history and yield immense fortunes for its owners, but it made important contributions to the development of the State of Nevada, as well as neighboring California. Included here are important details on not only the early development and history of the Comstock, but also rare early insight into its mines, ore and its geology.**8.5" X 11", 244 ppgs, Retail Price: $19.99**

Colorado Mining Books

<u>Ores of The Leadville Mining District</u> - Unavailable since 1926, this publication was originally compiled by the United States Department of Interior. This volume also includes important insights into the ores and mineralization of the Leadville Mining District in Colorado. Topics include historic ore prospecting methods, local geology, insights into ore veins and stockworks, the local trend and distribution of ore channels, reverse faults, shattered rock above replacement ore bodies, mineral enrichment in oxidized and sulphide zones and more. **8.5" X 11", 66 ppgs, Retail Price: $8.99**

<u>Mining in Colorado</u> - Unavailable since 1926, this publication was originally compiled by the United States Department of Interior. This volume also includes important insights into the mining history of Colorado from its early beginnings in the 1850's right up to the mid 1920's. Not only is Colorado's gold mining heritage included, but also its silver, copper, lead and zinc mining industry. Each mining area is treated separately, detailing the development of Colorado's mines on a county by county basis. **8.5" X 11", 284 ppgs, Retail Price: $19.99**

<u>Gold Mining in Gilpin County Colorado</u> - Unavailable since 1876, this publication was originally compiled by the Register Steam Printing House of Central City, Colorado. A rare glimpse at the gold mining history and early mines of Gilpin County, Colorado from their first discovery in the 1850's up to the "flush years" of the mid 1870's. Of particular interest is the history of the discovery of gold in Gilpin County and details about the men who made those first strikes. Special focus is given to the early gold mines and first mining districts of the area, many of which are not detailed in other books on Colorado's gold mining history. **8.5" X 11", 156 ppgs, Retail Price: $12.99**

<u>Mining in the Gold Brick Mining District of Colorado</u> - Important insights into the history of the Gold Brick Mining District, as well as its local geography and economic geology. Also included are the histories and locations of historic mines in this important Colorado Mining District, including the Cortland, Carter, Raymond, Gold Links, Sacramento, Bassick, Sandy Hook, Chronicle, Grand Prize, Chloride, Granite Mountain, Lucille, Gray Mountain, Hilltop, Maggie Mitchell, Silver Islet, Revenue, Roosevelt, Carbonate King and others. In addition to hardrock mining, are also included are details on gold placer mining in this portion of Colorado. **8.5" X 11", 140 ppgs, Retail Price: $12.99**

Washington Mining Books

<u>The Republic Mining District of Washington</u> - Unavailable since 1910, this important publication was originally published by the Washington Geologic Survey and has been unavailable for a century. Topics include the geology, rock formations and the formation of ore deposits in this important mining area of Washington State. Also included are hard to find details on the geology, history and locations of dozens of mines in the area. Some of the mines featured include the New Republic Mine, Ben Hur, Morning Glory, the South Republic Mine, Quilp, Surprise, Black Tail, Lone Pine, San Poil, Mountain Lion, Tom Thumb, Elcaliph and many others. **8.5" X 11", 94 ppgs, Retail Price: $10.99**

The Myers Creek and Nighthawk Mining Districts of Washington - Unavailable since 1911, this important publication was originally published by the Washington Geologic Survey and has been unavailable for a century. Topics include the geology, rock formations and the formation of ore deposits in these important mining areas of Washington State. Also included are hard to find details on the geology, history and locations of dozens of mines in the area. Some of the mines featured include the Grant Mine, Monterey, Nip and Tuck, Myers Creek, Number Nine, Neutral, Rainbow, Aztec, Crystal Butte, Apex, Butcher Boy, Molson, Mad River, Olentangy, Delate, Kelsey, Golden Chariot, Okanogan, Ohio, Forty-Ninth Parallel, Nighthawk, Favorite, Little Chopaka, Summit, Number One, California, Peerless, Caaba, Prize Group, Ruby, Mountain Sheep, Golden Zone, Rich Bar, Similkameen, Kimberly, Triune, Hiawatha, Trinity, Hornsilver, Maquae, Bellevue, Bullfrog, Palmer Lake, Ivanhoe, Copper World and many others.
8.5" X 11", 136 ppgs, Retail Price: $12.99

The Blewett Mining District of Washington - Unavailable since 1911, this important publication was originally published by the Washington Geologic Survey and has been unavailable for a century. Topics include the geology, rock formations and the formation of ore deposits in this important mining area of Washington State. Also included are hard to find details on the geology, history and locations of dozens of mines in the area. Some of the mines featured include the Washington Meteor, Alta Vista, Pole Pick, Blinn, North Star, Golden Eagle, Tip Top, Wilder, Golden Guinea, Lucky Queen, Blue Bell, Prospect, Homestake, Lone Rock, Johnson, and others. **8.5" X 11", 134 ppgs, Retail Price: $12.99**

Silver Mining In Washington - Unavailable since 1955, this important publication was originally published by the Washington Geologic Survey. Featured are the hard to find locations and details pertaining to Washington's silver mines. **8.5" X 11", 180 ppgs, Retail Price: $15.99**

The Mines of Snohomish County Washington - Unavailable since 1942, this important publication was originally published by the Washington Geologic Survey and has been unavailable for seventy years. Featured are details on a large number of gold, silver, copper, lead and other metallic mineral mines. Included are the locations of each historic mine, along with information on the commodity produced. **8.5" X 11", 98 ppgs, Retail Price: $10.99**

The Mines of Chelan County Washington - Unavailable since 1943, this important publication was originally published by the Washington Geologic Survey and has been unavailable for seventy years. Featured are details on a large number of gold, silver, copper, lead and other metallic mineral mines. Included are the locations of each historic mine, along with information on the commodity. **8.5" X 11", 88 ppgs, Retail Price: $9.99**

Metal Mines of Washington - Unavailable since 1921, this important publication was originally published by the Washington Geologic Survey and has been unavailable for nearly ninety years. Widely considered a masterpiece on the Washington Mining Industry, "Metal Mines of Washington" sheds light on the important details of Washington's early mining years. Featured are details on hundreds of gold, silver, copper, lead and other metallic mineral mines. Included are hard to find details on the mineral resources of this state, as well as the locations of historic mines. Lavishly illustrated with maps and historic photos and complete with a glossary to explain any technical terms found in the text, this is one of the most important works on mining in the State of Washington. No prospector or miner should be without it if they are interested in mining in Washington. **8.5" X 11", 396 ppgs, Retail Price: $24.99**

Gem Stones In Washington - Unavailable since 1949, this important publication was originally published by the Washington Geologic Survey and has been unavailable since first published. Included are details on where to find naturally occurring gem stones in the State of Washington, including quartz crystal, amethyst, smoky quartz, milky quartz, agates, bloodstone, carnelian, chert, flint, jasper, onyx, petrified wood, opal, fire opal, hyalite and others. **8.5" X 11", 54 ppgs, Retail Price: $8.99**

The Covada Mining District of Washington - Unavailable since 1913, this important publication was originally published by the Washington Geologic Survey and has been unavailable for a century. Topics include the geology, rock formations and the formation of ore deposits in this important mining area of Washington State. Also included are hard to find details on the geology, history and locations of dozens of mines in the area. Some of the mines featured include the Admiral, Advance, Algonkian, Big Bug, Big Chief, Big Joker, Black Hawk, Black Tail, Black Thorn, Captain, Cherokee Strip, Colorado, Dan Patch, Dead Shot, Etta, Good Ore, Greasy Run, Great Scott, Idora, IXL, Jay Bird, Kentucky Bell, King Solomon, Laurel, Laura S, Little Jay, Meteor, Neglected, Northern Light, Old Nell, Plymouth Rock, Polaris, Quandary, Reserve, Shoo Fly, Silver Plume, Three Pines, Vernie, White Rose and dozens of others. **8.5" X 11", 114 ppgs, Retail Price: $10.99**

The Index Mining District of Washington - Unavailable since 1912, this important publication was originally published by the Washington Geologic Survey and has been unavailable for a century. Topics include the geology, rock formations and the formation of ore deposits in this important mining area of Washington State. Also included are hard to find details on the geology, history and locations of dozens of mines in the area. Some of the mines featured include the Sunset, Non-Pareil, Ethel Consolidated, Kittaning, Merchant, Homestead, Co-operative, Lost Creek, Uncle Sam, Calumet, Florence-Rae, Bitter Creek, Index Peacock, Gunn Peak, Helena, North Star, Buckeye. Copper Bell, Red Cross and others. **8.5" X 11", 114 ppgs, Retail Price: $11.99**

Mining & Mineral Resources of Stevens County Washington - Unavailable since 1920, this important publication was originally published by the Washington Geologic Survey and has been unavailable for a century. Topics include the geology, rock formations and the formation of ore deposits in these important mining areas of Washington State. Also included are hard to find details on the geology, history and locations of hundreds of mines in the area. **8.5" X 11", 372 ppgs, Retail Price: $24.99**

The Mines and Geology of the Loomis Quadrangle Okanogan County, Washington - Unavailable since 1972, this important publication was originally published by the Washington Geologic Survey and has been unavailable for a century. Topics include the geology, rock formations and the formation of ore deposits in this important mining area of Washington State. Also included are hard to find details on the geology, history and locations of dozens of gold, copper, silver and other mines in the area. **8.5" X 11", 150 ppgs, Retail Price: $12.99**

The Conconully Mining District of Okanogan County Washington - Unavailable since 1973, this important publication was originally published by the Washington Geologic Survey and has been unavailable for a century. Topics include the geology, rock formations and the formation of ore deposits in this important mining area of Washington State, which also includes Salmon Creek, Blue Lake and Galena. Also included are hard to find details on the geology, mining history and locations of dozens of mines in the area. Some of the mines include Arlington, Fourth of July, Sonny Boy, First Thought, Last Chance, War Eagle-Peacock, Wheeler, Mohawk, Lone Star, Woo Loo Moo Loo, Keystone, Hughes, Plant-Callahan, Johnny Boy, Leuena, Gubser, John Arthur, Tough Nut, Homestake, Key and many others **8.5" X 11", 68 ppgs, Retail Price: $8.99**

Wyoming Mining Books

Mining in the Laramie Basin of Wyoming - Unavailable since 1909, this publication was originally compiled by the United States Department of Interior. Also included are insights into the mineralization and other characteristics of this important mining region, especially in regards to coal, limestone, gypsum, bentonite clay, cement, sand, clay and copper. **8.5" X 11", 104 ppgs, Retail Price: $11.99**

New Mexico Mining Books

The Mogollon Mining District of New Mexico - Unavailable since 1927, this important publication was originally published by the US Department of Interior and has been unavailable for 80 years. Topics include the geology, rock formations and the formation of ore deposits in this important mining area in New Mexico. Of particular focus is information on the history and production of the ore deposits in this area, their form and structure, vein filling, their paragenesis, origins and ore shoots, as well as oxidation and supergene enrichment. Also included are hard to find details, including the descriptions and locations of numerous gold, silver and other types of mines, including the Eureka, Pacific, South Alpine, Great Western, Enterprise, Buffalo, Mountain View, Floride, Gold Dust, Last Chance, Deadwood, Confidence, Maud S., Deep Down, Little Fanney, Trilby, Johnson, Alberta, Comet, Golden Eagle, Cooney, Queen, the Iron Crown, Eberle, Clifton, Andrew Jackson mine, Mascot and others. **8.5" X 11", 144 ppgs, Retail Price: $12.99**

The Percha Mining District of Kingston New Mexico - Unavailable since 1883, this important publication was originally published by the Kingston Tribune and has been unavailable for over one hundred and thirty five years. Having been written during the earliest years of gold and silver mining in the Percha Mining District, unlike other books on the subject, this work offers the unique perspective of having actually been written while the early mining history of this area was still being made. In fact, the work was written so early in the development of this area that many of the notable mines in the Percha District were less than a few years old and were still being operated by their original discoverers with the same enthusiasm as when they were first located. Included are hard to find details on the very earliest gold and silver mines of this important mining district near Kingston in Sierra County, New Mexico. **8.5" X 11", 68 ppgs, Retail Price: $9.99**

East Coast Mining Books

The Gold Fields of the Southern Appalachians - Unavailable since 1895, this important publication was originally published by the US Department of Interior and has been unavailable for nearly 120 years. Topics include the geology, rock formations and the formation of ore deposits in this important mining area of the American South. Of particular focus is information on the history and statistics of the ore deposits in this area, their form and structure and veins. Also included are details on the placer gold deposits of the region. The gold fields of the Georgian Belt, Carolinian Belt and the South Mountain Mining District of North Carolina are all treated in descriptive detail. Included are hard to find details, including the descriptions and locations of numerous gold mines in Georgia, North Carolina and elsewhere in the American South. Also included are details on the gold belts of the British Maritime Provinces and the Green Mountains. **8.5" X 11", 104 ppgs, Retail Price: $9.99**

Gold Rush Tales Series

Millions in Siskiyou County Gold - In this first volume of the "Gold Rush Tales" series, leading mining historian and editor Kerby Jackson, introduces us to the story of how millions of dollars worth of gold was discovered in Siskiyou County during the California Gold Rush. Lavishly illustrated with photos from the 19th Century, this hard to find information was first published in 1897 and sheds important light onto the gold rush era in Siskiyou County, California and the experiences of the men who dug for the gold and actually found it. **8.5" X 11", 82 ppgs, Retail Price: $9.99**

The California Rand in the Days of '49 - In this second volume of the "Gold Rush Tales" series, leading mining historian and editor Kerby Jackson, introduces us to four tales from the California Gold Rush. Lavishly illustrated with photos from the 19th Century, this hard to find information was first published in 1890's and includes the stories of "California's Rand", details about Chinese miners, how one early miner named Baker struck it rich and also the story of Alphonzo Bowers, who invented the first hydraulic gold dredge. **8.5" X 11", 54 ppgs, Retail Price: $9.99**

More Mining Books

Prospecting and Developing A Small Mine - Topics covered include the classification of varying ores, how to take a proper ore sample, the proper reduction of ore samples, alluvial sampling, how to understand geology as it is applied to prospecting and mining, prospecting procedures, methods of ore treatment, the application of drilling and blasting in a small mine and other topics that the small scale miner will find of benefit. **8.5" X 11", 112 ppgs, Retail Price: $11.99**

Timbering For Small Underground Mines - Topics covered include the selection of caps and posts, the treatment of mine timbers, how to install mine timbers, repairing damaged timbers, use of drift supports, headboards, squeeze sets, ore chute construction, mine cribbing, square set timbering methods, the use of steel and concrete sets and other topics that the small underground miner will find of benefit. This volume also includes twenty eight illustrations depicting the proper construction of mine timbering and support systems that greatly enhance the practical usability of the information contained in this small book. **8.5" X 11", 88 ppgs. Retail Price: $10.99**

Timbering and Mining - A classic mining publication on Hard Rock Mining by W.H. Storms. Unavailable since 1909, this rare publication provides an in depth look at American methods of underground mine timbering and mining methods. Topics include the selection and preservation of mine timbers, drifting and drift sets, driving in running ground, structural steel in mine workings, timbering drifts in gravel mines, timbering methods for driving shafts, positioning drill holes in shafts, timbering stations at shafts, drainage, mining large ore bodies by means of open cuts or by the "Glory Hole" system, stoping out ore in flat or low lying veins, use of the "Caving System", stoping in swelling ground, how to stope out large ore bodies, Square Set timbering on the Comstock and its modifications by California miners, the construction of ore chutes, stoping ore bodies by use of the "Block System", how to work dangerous ground, information on the "Delprat System" of stoping without mine timbers, construction and use of headframes and much more. This volume provides a reference into not only practical methods of mining and timbering that may be employed in narrow vein mining by small miners today, but also rare insights into how mines were being worked at the turn of the 19th Century. **8.5" X 11", 288 ppgs. Retail Price: $24.99**

A Study of Ore Deposits For The Practical Miner - Mining historian Kerby Jackson introduces us to a classic mining publication on ore deposits by J.P. Wallace. First published in 1908, it has been unavailable for over a century. Included are important insights into the properties of minerals and their identification, on the occurrence and origin of gold, on gold alloys, insights into gold bearing sulfides such as pyrites and arsenopyrites, on gold bearing vanadium, gold and silver tellurides, lead and mercury tellurides, on silver ores, platinum and iridium, mercury ores, copper ores, lead ores, zinc ores, iron ores, chromium ores, manganese ores, nickel ores, tin ores, tungsten ores and others. Also included are facts regarding rock forming minerals, their composition and occurrences, on igneous, sedimentary, metamorphic and intrusive rocks, as well as how they are geologically disturbed by dikes, flows and faults, as well as the effects of these geologic actions and why they are important to the miner. Written specifically with the common miner and prospector in mind, the book will help to unlock the earth's hidden wealth for you and is written in a simple and concise language that anyone can understand. **8.5" X 11", 366 ppgs. Retail Price: $24.99**

Mine Drainage - Unavailable since 1896, this rare publication provides an in depth look at American methods of underground mine drainage and mining pump systems. This volume provides a reference into not only practical methods of mining drainage that may be employed in narrow vein mining by small miners today, but also rare insights into how mines were being worked at the turn of the 19th Century. **8.5" X 11", 218 ppgs. Retail Price: $24.99**

Fire Assaying Gold, Silver and Lead Ores - Unavailable since 1907, this important publication was originally published by the Mining and Scientific Press and was designed to introduce miners and prospectors of gold, silver and lead to the art of fire assaying. Topics include the fire assaying of ores and products containing gold, silver and lead; the sampling and preparation of ore for an assay; care of the assay office, assay furnaces; crucibles and scorifiers; assay balances; metallic ores; scorification assays; cupelling; parting' crucible assays, the roasting of ores and more. This classic provides a time honored method of assaying put forward in a clear, concise and easy to understand language that will make it a benefit to even beginners. **8.5" X 11", 96 ppgs. Retail Price: $11.99**

Methods of Mine Timbering - Originally published in 1896, this important publication on mining engineering has not been available for nearly a century. Included are rare insights into historical methods of timbering structural support that were used in underground metal mines during the California that still have a practical application for the small scale hardrock miner of today. **8.5" X 11", 94 ppgs. Retail Price: $10.99**

The Enrichment of Copper Sulfide Ores - First published in 1913, it has been unavailable for over a century. Topics include the definition and types of ore enrichment, the oxidation of copper ores, the precipitation of metallic sulfides. Also included are the results of dozens of lab experiments pertaining to the enrichment of sulfide ores that will be of interest to the practical hard rock mine operator in his efforts to release the metallic bounty from his mine's ore. **8.5" X 11", 92 ppgs. Retail Price: $9.99**

A Study of Magmatic Sulfide Ores - Unavailable since 1914, this rare publication provides an in depth look at magmatic sulfide ores. Some of the topics included are the definition and classification of magmatic ores, descriptions of some magmatic sulfide ore deposits known at the time of publication including copper and nickel bearing pyrrohitic ore bodies, chalcopyrite-bornite deposits, pyritic deposits, magnetite-ileminite deposits, chromite deposits and magmatic iron ore deposits. Also included are details on how to recognize these types of ore deposits while prospecting for valuable hardrock minerals. **8.5" X 11", 138 ppgs. Retail Price: $11.99**

The Cyanide Process of Gold Recovery - Unavailable since 1894 and released under the name "The Cyanide Process: Its Practical Application and Economical Results", this rare publication provides an in depth look at the early use of cyanide leaching for gold recovery from hardrock mine ores. This volume provides a reference into the early development and use of cyanide leaching to recover gold. **8.5" X 11", 162 ppgs. Retail Price: $14.99**

California Gold Milling Practices - Unavailable since 1895 and released under the name "California Gold Practices", this rare publication provides an in depth look at early methods of milling used to reduce gold ores in California during the late 19th century. This volume provides a reference into the early development and use of milling equipment during the earliest years of the California Gold Rush up to the age of the Industrial Revolution. Much of the information still applies today and will be of use to small scale miners engaging in hardrock mining. **8.5" X 11", 104 ppgs. Retail Price: $10.99**

Leaching Gold and Silver Ores With The Plattner and Kiss Processes - Mining historian Kerby Jackson introduces us to a classic mining publication on the evaluation and examination of mines and prospects by C.H. Aaron. First published in 1881, it has been unavailable for over a century and sheds important light on the leaching of gold and silver ores with the Plattner and Kiss processes. **8.5" X 11", 204 ppgs. Retail Price: $15.99**

The Metallurgy of Lead and the Desilverization of Base Bullion - First published in 1896, it has been unavailable for over a century and sheds important light on the the recovery of silver from lead based ores. Some of the topics include the properties of lead and some of its compounds, lead ores such as galenite, anglesite, cerussite and others, the distribution of lead ores throughout the United States and the sampling and assaying of lead ores. Also covered is the metallurgical treatment of lead ores, as well as the desilverization of lead by the Pattinson Process and the Parkes Process. Hofman's text has long been considered one of the most important early works on the recovery of silver from lead based ores. **8.5" X 11", 452 ppgs. Retail Price: $29.99**

Ore Sampling For Small Scale Miners - First published in 1916, it has been unavailable for over a century and sheds important light on historic methods of ore sampling in hardrock mines. Topics include how to take correct ore samples and the conditions that affect sampling, such as their subdivision and uniformity. Particular detail is given to methods of hand sampling ore bodies by grab sample, pipe sample and coning, as well as sampling by mechanical methods. Also given are insights into the screening, drying and grinding processes to achieve the most consistent sample results and much more. **8.5" X 11", 124 ppgs. Retail Price: $12.99**

The Extraction of Silver, Copper and Tin from Ores - First published in 1896, it has been unavailable for over a century and sheds important light on how historic miners recovered silver, copper and tin from their mining operations. The book is split into three sections, including a discussion on the Lixiviation of Silver Ores, the mining and treatment of copper ores as practiced at Tharsis, Spain and the smelting of tin as it was practiced by metallurgists at Pulo Brani, Singapore. Also included is an overview and analysis of these historic metal recovery methods that will be of benefit to those interested in the extraction of silver, copper and tin from small mines. **8.5" X 11", 118 ppgs. Retail Price: $14.99**

The Roasting of Gold and Silver Ores - First published in 1880, it has been unavailable for over a century and sheds important light on how historic miners recovered gold and silver rom their mining operations. Topics include details on the most important silver and free milling gold ores, methods of desulphurization of ores, methods of deoxidation, the chlorination of ores, methods and details on roasting gold and silver ores, notes on furnaces and more. Also included are details on numerous methods of gold and silver recovery, including the Ottokar Hofman's Process, the Patera Process, Kiss Process, Augustin Process, Ziervogel Process and others. **8.5" X 11", 178 ppgs. Retail Price: $19.99**

The Examination of Mines and Prospects - First published in 1912, it has been unavailable for over a century and sheds important light on how to examine and evaluate hardrock mines, prospects and lode mining claims. Sections include Mining Examinations, Structural Geology, Structural Features of Ore Deposits, Primary Ores and their Distribution, Types of Primary Ore Deposits, Primary Ore Shoots, The Primary Alteration of Wall Rocks, Alterations by Surface Agencies, Residual Ores and their Distribution, Secondary Ores and Ore Shoots and Vein Outcrops. This hard to find information is a must for those who are interested in owning a mine or who already own a lode mining claim and wish to succeed at quartz mining. **8.5" X 11", 250 ppgs. Retail Price: $19.99**